Sustainable Development Goals Series

The **Sustainable Development Goals Series** is Springer Nature's inaugural cross-imprint book series that addresses and supports the United Nations' seventeen Sustainable Development Goals. The series fosters comprehensive research focused on these global targets and endeavours to address some of society's greatest grand challenges. The SDGs are inherently multidisciplinary, and they bring people working across different fields together and working towards a common goal. In this spirit, the Sustainable Development Goals series is the first at Springer Nature to publish books under both the Springer and Palgrave Macmillan imprints, bringing the strengths of our imprints together.

The Sustainable Development Goals Series is organized into eighteen subseries: one subseries based around each of the seventeen respective Sustainable Development Goals, and an eighteenth subseries, "Connecting the Goals", which serves as a home for volumes addressing multiple goals or studying the SDGs as a whole. Each subseries is guided by an expert Subseries Advisor with years or decades of experience studying and addressing core components of their respective Goal.

The SDG Series has a remit as broad as the SDGs themselves, and contributions are welcome from scientists, academics, policymakers, and researchers working in fields related to any of the seventeen goals. If you are interested in contributing a monograph or curated volume to the series, please contact the Publishers: Zachary Romano [Springer; zachary.romano@springer.com] and Rachael Ballard [Palgrave Macmillan; rachael.ballard@palgrave.com].

Huadong Guo

Big Earth Data in Support of the Sustainable Development Goals (2023)—The Belt and Road

Science Press
Beijing

Springer

Huadong Guo
International Research Center of Big
Data for Sustainable Development
Goals (CBAS)
Chinese Academy of Sciences
Aerospace Information Research
Institute
Beijing, China

ISSN 2523-3084 ISSN 2523-3092 (electronic)
Sustainable Development Goals Series
ISBN 978-981-95-3177-6 ISBN 978-981-95-3178-3 (eBook)
https://doi.org/10.1007/978-981-95-3178-3

Jointly published with Science Press.
The print edition is not for sale in China (Mainland). Customers from China (Mainland) please order the print book from: Science Press.
ISBN of the Co-Publisher's edition: 978-7-03-080887-5

This work was supported by Chinese Academy of Sciences.

This Springer imprint is published by the registered company Springer Nature Singapore Pte Ltd.
The registered company address is: 152 Beach Road, #21-01/04 Gateway East, Singapore 189721, Singapore

Editorial Board

Principal Authors

Preface

From 2015 to the present, the implementation of the United Nations (UN) *Transforming Our World: The 2030 Agenda for Sustainable Development* (referred to as the 2030 Agenda) has reached the halfway point. During this period, progress on the Sustainable Development Goals (SDGs) has encountered serious challenges and difficulties despite the overall advancements. The international community has gained deeper insights into long-term issues such as climate change and biodiversity loss, but further efforts are needed to enhance implementation. Unexpected events like the COVID-19 pandemic and regional conflicts have had significant and far-reaching impacts on society, the economy, and the environment.

It is important for the international community to learn both the experience and lessons of the first half of the 2030 Agenda, so that we can enhance the implementation of the SDGs in the second half and explore directions for future sustainable development. The slow progress during the initial phase underscores the global inadequacy in addressing long-term risks as well as short-term crises in the process of promoting sustainable development. The lack of timely and accurate data remains a weakness in our response to both long-term and short-term issues, and it is a bottleneck hindering the implementation and monitoring of the SDGs and the formulation of science-based decisions.

In September 2021, Chinese President Jinping Xi proposed the Global Development Initiative (GDI) during the 76th Session of The UN General Assembly. The GDI aims to deepen international cooperation, accelerate the implementation of the 2030 Agenda, and promote stronger, greener, and healthier development. It places a special emphasis on advancing collaboration in areas such as digital connectivity in the digital age to expedite the realization of the 2030 Agenda. The GDI is committed to building a global community of development, and a crucial foundation in this endeavor is the promotion of digital technology. Big Earth Data, as a representative form of such technology, can play a vital role in filling the gaps in current SDG statistical data and spatiotemporal information.

The year 2023 marks the third anniversary of President Jinping Xi's announcement of the establishment of the International Research Center of Big Data for Sustainable Development Goals (CBAS) in China, and also the fifth consecutive year of publishing the report Big Earth Data in Support of the Sustainable Development Goals (referred to as the report). Over the past few years, the research team has made use of the increasingly refined

Big Earth Data platform and leveraged the advantages of objective scientific data and rich spatiotemporal information to assess the midterm progress on sustainable development in China and globally. The team's focus has been on exploring the bottlenecks hindering sustainable development and identifying future development directions.

This year's report focuses on the midterm evaluation of sustainable development in China and globally through Big Earth Data. It aims to further expand data products for SDG indicators, broaden the depth of indicator contents, complete progress evaluations of all environment-related SDG indicators in China, and actively provide public data products to serve all the countries, particularly developing countries in terms of evaluating the implementation of the 2030 Agenda. It also provides recommendations on future science-based decision-making, accelerating sustainable development processes, building big data acquisition capabilities, and adjusting and optimizing indicators. The report stands as scientific support for the implementation of the 2030 Agenda.

The report is a collaborative effort, written by over 160 researchers from more than 50 organizations, including research institutes and universities in the field of sustainable development and big data. The project has received strong support from the Chinese Academy of Sciences and various government departments, and the team members have dedicated significant efforts to this endeavor. We express our heartfelt gratitude to all those involved in making this report a reality.

Huadong Guo
Beijing, China

Director-General of the International
Research Center of Big Data
for Sustainable Development Goals

Former Member, the 10-Member group
of the United Nations Sustainable
Development Goals Technology Facilitation Mechanism

Member, International Science Council Global Commission
on Science Missions for Sustainability

Competing Interests The author has no competing interests to declare that are relevant to the content of this manuscript.

Executive Summary

This report gives full play to the advantages and characteristics of Big Earth Data. It focuses on 19 targets related to seven SDGs, including SDG 2 (Zero Hunger), SDG 6 (Clean Water and Sanitation), SDG 7 (Affordable and Clean Energy), SDG 11 (Sustainable Cities and Communities), SDG 13 (Climate Action), SDG 14 (Life Below Water), SDG 15 (Life on Land), as well as the interactions among SDG indicators, resulting in 25 research cases. From three aspects: data products, methods and models, and decision support, this report shows the results of monitoring and midterm evaluation of relevant SDG indicators at four scales: global, regional, national, and typical areas. This report represents an innovative practice of big data in support of SDG implementation and can provide scientific reference for decision-makers.

Regarding Zero Hunger (SDG 2), this report assesses the global food production potential, with a particular focus on Southeast Asia, a region that is highly sensitive to food security issues. The study specifically monitors the cultivation of rice, which is the primary staple crop in the region. By utilizing machine learning, a cross-entropy distribution probability model, time-series statistical features, and a U-Net model for large-scale tropical/subtropical SAR-based annual rice mapping, several datasets such as a dataset of yield gaps for major global crops from 2010 to 2020 at five-year intervals, a spatial distribution dataset of multiple crop types' cultivation areas and yields in Southeast Asian countries for 2020, and an interannual rice cultivation area extraction dataset for five Southeast Asian countries from 2019 to 2021 were generated. The key factors influencing per unit area yield potential of global crops and rice cultivation areas in Southeast Asia have been identified. The trends in total and per capita crop production in Southeast Asian countries have been summarized, and recommendations for rice cultivation methods and context-specific food security policies in the region have been provided.

Regarding Clean Water and Sanitation (SDG 6), this report focuses on three specific targets: improving water quality (SDG 6.3), enhancing water-use efficiency (WUE) (SDG 6.4), and protecting and restoring water-related ecosystems (SDG 6.6). Based on Big Earth Data technology, this report presents three global-scale case studies. These include the creation of the global large lake and reservoir water transparency spatial distribution dataset for 2000–2021, the global farmland WUE dataset for 2001–2020, and the triennial global water occurrence frequency dataset for 2001–2021. Using these datasets, this report analyzes the changes in lake and reservoir water transparency, farmland

WUE, and the areas of natural and artificial water bodies at the global, continental, regional, and national scales. This report demonstrates the feasibility and advantages of using Big Earth Data technology for monitoring and evaluating changes in surface water environments, water resource use efficiency, and water ecosystem conditions.

Regarding Affordable and Clean Energy (SDG7), this report focuses on four specific targets and four indicators, and conducts a midterm progress evaluation of China. The case studies find that significant progress has been made in all four Chinese SDG 7 indicators evaluated. China has enhanced Africa's ability to obtain electricity through investment and financing, project construction, product supply, technology transfer, personnel training, and planning. China's overseas parks actively promote the experience of green and low-carbon development and promote the green and low-carbon development of local energy utilization.

Regarding Sustainable Cities and Communities (SDG 11), this report focuses on six indicators and assesses the progress of SDG 11 indicators on a global scale. The study found that from 2015 to 2020, the proportion of the population with convenient access to public transportation in major cities around the world (SDG 11.2.1) increased by 3.38%. In the past decade or so, the urban areas of Global South Countries and Global North Countries worldwide developed rapidly. In particular, the growth rate of urban areas in China was almost twice that of Australia and the United States. The global situation of heritage protection has shown a positive trend. Since the implementation of the Sendai Framework for Disaster Risk Reduction 2015–2030, both the annual average death toll indicator and the direct economic loss indicator caused by extreme weather and climate events worldwide have shown obvious regional differences. Across the globe, the increase in greenery in the built-up areas of cities in developing countries is the most significant, with China contributing the most to urban greening. The sources of atmospheric particulate matter in typical cities around the world are affected by multiple factors. These studies provide important support for the refined analysis of the progress of indicators in different regions and for scientific decision-making regarding sustainable cities and communities.

Regarding Climate Action (SDG 13), this report focuses on SDG 13.2.2 (greenhouse gas emissions and global ecosystem source-sink assessments). Research has found that global greenhouse gas emissions still face significant pressure. After a brief decline in 2020, global greenhouse gas emissions returned to an upward trajectory after 2021, and global CO_2 emissions increased by 1.5% in 2022 compared to 2021; From 1980 to 2022, global terrestrial ecosystems showed a significant CO_2 carbon sink with an average intensity of 2.36 Pg C/a. For CH_4 and N_2O, the global terrestrial ecosystems were a source of atmospheric emissions, with an average emission intensity of 173 Tg C/a and 8.9 Tg N/a from 1980 to 2022, respectively. The long-term trends of the three greenhouse gases also showed significant differences, with carbon sinks showing a significant increase from 1980 to 2000. However, since 2000, the increasing trend of global terrestrial ecosystem carbon sinks has stagnated. CH_4 emissions showed a trend of first increasing and then decreasing, with a turning point in the trend around 2012. N_2O emissions significantly increased from 1980 to 2016 and began to show a decreasing trend after 2016.

Regarding Life Below Water (SDG 14), focusing on the themes of SDG 14.1 to reduce marine pollution and SDG 14.2 to protect marine ecosystems, three typical case studies were conducted: the distribution and migration characteristics of microplastics in the Arctic-Pacific sector, the spatiotemporal variations of large-scale phytoplankton blooms in the North Indian Ocean, and the dynamic monitoring of live coral cover on typical islands and reefs. The studies found that there is diversity in the types of microplastics present in the water bodies, sediments, and organisms in the Arctic-Pacific sector, and that microplastic aggregation can occur in areas where eddies and ocean currents converge in the Arctic Ocean. The Northern Arabian Sea is the most significant area for annual large-scale phytoplankton blooms in the North Indian Ocean. Besides nutrients, sea surface temperature is the primary factor influencing the changes in large-scale phytoplankton blooms in most of the North Indian Ocean. Dynamic monitoring results of live coral cover using multi-spectral satellite remote sensing in 2015, 2020, and 2022 at the Hawaii Reef (Kure Reef) and the Great Barrier Reef (Nameless Reef) indicate that the live coral cover on islands and reefs in these two typical regions is decreasing overall. The monitoring methods, data products, and recommendations related to these indicators can provide important references for better achieving SDG 14.

Regarding Life on Land (SDG 15), this report focuses on two specific targets and conducts assessments on global and regional scales. The findings show that in combating desertification, the processes of grassland degradation and restoration occurred simultaneously in different regions worldwide between 2000 and 2020. However, the area of restored grasslands was approximately 1.9 times larger than that of degraded grasslands, indicating a clear trend of grassland recovery globally, with over 40% of the restored grasslands showing significant improvement. Both climate and human factors were identified as key drivers in the processes of grassland degradation and restoration. On the Mongolian Plateau, the total area of desertified land decreased from 1,698,000 km^2 in 2015 to 1,656,400 km^2 in 2020. In terms of natural habitat protection, migratory bird richness in Los Angeles, Mumbai, and Kuala Lumpur was relatively high, with values of 66, 61, and 58, respectively. In contrast, Cairo and Bandung showed relatively low richness, with values of 8 and 18, respectively. Furthermore, from 1992 to 2022, nighttime light intensity in major cities worldwide generally exhibited an expanding trend.

Regarding SDGs' integrated assessment and interactions, the report analyzes the future scenarios, regional coordination, and balanced development of global SDGs under the interaction of multiple SDGs. The research findings indicate significant spatial disparities in the global progress toward SDGs. By 2030, most countries are expected to meet the targets for basic needs SDGs, while environmental protection indicators face varying degrees of challenges. Some developing countries need to accelerate their economic development, while certain middle- and high-income countries should shift towards low-carbon consumption and production models to promote sustainable environmental development. From 2010 to 2020, significant negative correlations were observed between the annual water yield and habitat quality, carbon storage and sequestration and habitat quality, as well as habitat quality and building volume, and between habitat quality and population density—building area in global waterfront areas. The relationship between global riparian ecosystem services and building volume shifted from a trade-off between 2010 and 2015 to a synergistic relationship from 2016 to 2020. Furthermore, the Belt and Road Initiative (BRI) has contributed a "Chinese solution" to addressing regional imbalances, with its contribution to reducing regional disparities exceeding 70%.

Contents

Acronyms

AFSIA	Africa Solar Industry Association
AFSIS	ASEAN Food Security Information System
AgMIP	Agricultural Model Intercomparison and Improvement Project
AR6	The 6th Assessment Report
BREEAM	Building research establishment environmental assessment method
BRI	Belt and Road Initiative
C3S	Copernicus Climate Change Service
CA	Cellular automata
CBAS	International Research Center of Big Data for Sustainable Development Goals
CFCD	China Forest Cover Dynamics
Chla	Chlorophyll a
CIE	Commission internationale del'éclairage chromaticity method
CMEMS	The Copernicus Marine Environment Monitoring Service
CMIP6	Commentary on the Coupled Model Intercomparison Project Phase 6
CP	Cellophane
CPE	Chlorinated polyethy-lene
CRU	Climatic Research Unit
DEM	Digital elevation model
DMSP	Defense Meteorological Satellite Program
DNB	Day/Night Band
DNN	Deep neural network
ECMWF	European Centre for Medium-Range Weather Forecasts
EDGAR	Emissions Database for Global Atmospheric Research
EF-LUE	Evaporative Fraction Light-Use-Efficiency
EM-DAT	Emergency Events Database
EMLS	European Multi Lake Survey
ENSO	El Nino and Southern oscillation
ERA5	Fifth-Generation ECMWF Atmospheric Reanalysis of the Global Climate
ESA	European Space Agency
ESA-CCI	European Space Agency-Climate Change Initiative
ET	Evapotranspiration

EVI	Enhanced vegetation index
FAO	Food and Agriculture Organization of the United Nations
GAEZ	Global Agro-Ecological Zones
GAIA	Global artificial impervious area
GDDAT	Global Disaster Data Platform
GDI	Global Development Initiative
GEE	Google Earth Engine
GEO	Group on Earth Observations
GGCMs	Global Gridded Crop Models
GHS-BUILT-V	Global Human Settlement-Built-up Volume grid
GHSL-UCDB	Global Human Settlement-Urban Centre Database
GHS-SMOD	Global Human Settlement-Settlement Model grid
GIS	Geographic information system
GLASS	Global land surface satellite
GPCC	Global Precipitation Climatology Centre
GPP	Gross primary productivity
GPS	Global positioning system
GRD	Ground Range Detected
GSW	Global surface water
GUB	Global urban boundaries
HIST	International Centre on Space Technologies for Natural and Cultural Heritage under the auspices of UNESCO
IAEG-SDGs	Inter-Agency and Expert Group on SDG Indicators
IBIS	Integrated Biosphere Simulator
ICOD	International Centre for Ocean Development
IGBP	International Geosphere-Biosphere Programme
InVEST	Integrated Valuation of Ecosystem Services and Trade-offs
IOD	Indian ocean dipole
IPCC	Intergovernmental Panel on Climate Change
IRENA	International Renewable Energy Agency
IRRI	International Rice Research Institute
ISA	Impervious surface area
ISIMIP	The Inter-Sectoral Impact Model Intercomparison Project
IUCN	International Union for Conservation of Nature
JAXA	Japan Aerospace Exploration Agency
JRC	Joint Research Centre
KDE	Kernel density estimation
LCC	Live coral cover
LCI	Low-Carbon Development Index
LCOE	Levelized cost of energy
LUH2	Land-Use Harmonization
MAE	Mean absolute error
MI	Moisture index
MIRCA	Global Monthly Irrigated and Rainfed Crop Areas
MNPP	Moisture-responded net primary productivity
MODIS	Moderate resolution imaging spectroradiometer
MPs	Microplastics

NASA	National Aeronautics and Space Administration
NDCs	Nationally determined contributions
nFLH	Normalized fluorescence line height
NPP	Net primary productivity
NTL	Nighttime light
OLI	Operational land imager
OLS	Operational Linescan System
OSM	OpenStreetMap
PA	Polyamide
PAA	Poly(acrylic acid)
PAM	Polyacrylamide
PAN	Polyacrylonitril
PAR	Photosynthetic active radiation
PCAPT	Public convenient accessibility to public transport
PE	Polyethylene
PET	Polyethylene terephthalate
POI	Point of interest
PP	Polypropylene
PS	Polystyrene
PTFE	Polytetrafluoroethylene
PVA	Poly(vinyl alcohol)
PVC	Polyvinyl chloride
RBD	Red band difference
RBI	Algal bloom index
RCP	Representative concentration pathway
RF	Random forest
RI	Red tide index
RMSE	Root mean square error
SAR	Synthetic aperture radar
SDGs	Sustainable Development Goals
SOA	State Oceanic Administration
SPAM	Spatial production allocation model
SRTM	Shuttle radar topography mission
SSP	Shared socioeconomic pathways
SST	Sea surface temperature
TROPOMI	TROPOspheric Monitoring Instrument
UNEP	United Nations Environment Programme
UNESCO	United Nations Educational, Scientific and Cultural Organization
UNFCCC	United Nations Framework Convention on Climate Change
VH	Vertical-horizontal
VIIRS	Visible infrared imaging radiometer suite
WAD2M	Wetland Area and Dynamics for Methane Modeling
WB	World Bank
WCRP	World Climate Research Programme

WDPA	World Database on Protected Areas
WHC	World Heritage Convention
WHO	World Health Organization
WRD	World Register of Dams
WUE	Water-use efficiency

Introduction

1

The year 2023 marks the midterm of the "2030 Agenda". It is imperative to assess the progress of current SDGs with more precise data to gain a comprehensive understanding of the challenges faced globally and by individual nations. This assessment will facilitate the exploration of scientifically feasible solutions and provide direction for the implementation phase of the agenda.

However, based on the analysis of the 17 goals in the UN's *Sustainable Development Goals Report 2022* and research results from over 230 indicators, it is evident that approximately half of the countries worldwide still severely lack progress data for these indicators. Additionally, the data updates are not timely, there is a lack of geospatial information, and these deficiencies significantly impede the monitoring of indicators and informed decision-making (United Nations 2022, 2023). The midterm evaluation of the SDGs presents an opportunity to assess the progress in implementing the SDGs and identify shortcomings, as well as to enhance data acquisition capabilities. According to the project team, out of the currently over 230 SDG indicators, 35% remain at Tier II, where methodologies exist but data are unavailable. Moreover, for those indicators with available data, their distribution across different countries and sectors is highly uneven. This is particularly challenging for many developing and underdeveloped countries, which find it difficult to achieve large-scale data computation in the short term and urgently require high-quality global data products.

The UN has promoted a "Digital Cooperation Roadmap" aimed at accelerating sustainable development through digital technologies (United Nations 2020). Big data is a crucial method and tool within digital technologies. In this report, we integrate Big Earth Data from multiple sources, including satellite observations, station records, survey statistics, online media, and foundational geographic data, to compensate for the absence of statistical data. This integration also provides rich spatiotemporal information, illustrating the spatial disparities and progress of indicators. The "Sustainable Development Science Satellite 1" (SDGSAT-1), launched in November 2021, is dedicated to serving the "2030 Agenda". Currently, satellite data are being shared globally to further enhance the acquisition and service capabilities of SDG data.

From 2019 to 2022, we have continuously published the "Big Earth Data in Support of the Sustainable Development Goals" (http://www.cbas.ac.cn/en/publications/reports/), aiding the realization of the SDGs in China and along the BRI countries. This report continues to focus on the progress of indicators at the scales of China and the BRI countries from the perspectives of data products, methods and models, and decision support. Additionally, this report emphasizes the midterm progress assessment

H. Guo, *Big Earth Data in Support of the Sustainable Development Goals (2023)—The Belt and Road*,
Sustainable Development Goals Series, https://doi.org/10.1007/978-981-95-3178-3_1

of indicators within China and the BRI countries. Among the more than 230 SDG indicators, over one-third are related to the environment (UNEP 2021). These environment-related indicators are also the areas where Big Earth Data's spatiotemporal advantages are most evident. As Earth Big Data research deepens, the data available for indicator computation are continuously increasing and being updated.

At the scale of BRI countries, we have primarily provided public data products from 2019 to 2023 related to SDG 2, SDG 6, SDG 7, SDG 11, SDG 13, SDG 14, and SDG 15. These include global arable land data products, global lake and reservoir water quality products, global electrification rate products, global urban impermeable surface and global urban public space products, global greenhouse gas emissions and natural disaster impact products, global marine aquaculture and mangrove distribution products, global forest cover and land degradation products, as well as SDGSAT-1 data products. These products are at the forefront in terms of resolution, timeliness and accuracy, and can directly support the assessment of global SDGs. Additionally, we have developed a Big Earth Data sharing platform and an online data visualization platform aimed at the sharing, display, and online computation of indicators.

Big data offers advantages such as rapid updates, repeatability, and extensive coverage in the computation of globally consistent indicators, playing a crucial role in addressing data deficiencies in certain countries (Guo 2020, 2021a, 2021b, 2022). However, during the long-term research and monitoring of indicators, from the perspective of big data analysis and computation, there remain three areas where the definitions and evaluation standards of some SDG indicators can be further refined and improved.

(1) Limited data comparability for certain indicators: Some indicators' data do not account for differences in population size, development levels, and geographical environments between countries. For instance, SDG 13.2.2, which measures annual greenhouse gas emissions, does not consider variations in population and economic scale across nations.
(2) Unclear evaluation standards for certain indicators: The criteria for some indicators lack quantifiability, making it challenging to assess progress and determine whether targets have been achieved. For example, SDG 14.a.1, which pertains to the proportion of the budget allocated to marine research activities, does not specify the threshold required to consider the goal as achieved. Furthermore, increasing the budget proportion for marine research may result in reduced allocations for other research areas such as agriculture and climate change.
(3) Difficulty in obtaining spatial data for certain indicators: Some indicators require differentiation based on gender, disability status, etc., such as SDG 11.2.1 (the proportion of the population with convenient access to public transportation) and SDG 11.7.1 (the proportion of urban public open spaces). While the intention may be to emphasize the protection of vulnerable populations, obtaining spatial distribution data for different genders and disabled populations within cities is challenging. Relying solely on statistical data fails to capture spatial disparities effectively.

In response to the current disparities in data acquisition capabilities among countries and the issues in indicator settings, we recommend that, on the one hand, countries strengthen support to continue enhancing their ability to compute indicators using Big Data. On the other hand, the UN should adjust the indicator settings by clarifying definitions and standards and increasing considerations for data accessibility. These measures will better leverage the role of globally consistent big data, advancing towards a framework that is "more equitable and accessible".

References

Guo HD (2020) Big earth data in support of the sustainable development goals (2019). Science Press and EDP Sciences, Beijing

Guo HD (2021a) Big earth data in support of the sustainable development goals (2020): the belt and road. Science Press and EDP Sciences, Beijing

Guo HD (2021b) Big earth data in support of the sustainable development goals (2020): China. Science Press and EDP Sciences, Beijing

Guo HD (2022) Big earth data in support of the sustainable development goals (2021): the belt and road. Science Press and EDP Sciences, Beijing

UNEP (2021) Measuring progress: environment and the SDGs. https://wedocs.unep.org/bitstream/handle/20.500.11822/35968/SDGMP.pdf

United Nations (2020) Road map for digital cooperation: implementation of the recommendations of the high-level panel on digital cooperation. https://digitallibrary.un.org/record/3864685/files/A_74_821-EN.pdf

United Nations (2022) The sustainable development goals report. https://desapublications.un.org/publications/sustainable-development-goals-report-2022

United Nations (2023) The sustainable development goals report 2023: special edition. https://unstats.un.org/sdgs/report/2023/The-Sustainable-Development-Goals-Report-2023.pdf

2 SDG 2 Zero Hunger

2.1 Background

More than halfway through the UN SDGs process, global progress towards achieving SDG 2 has been diverted by climate change, extreme weather events, political conflicts, economic fluctuations, and increasing social inequality (United Nations 2022). Projections indicate that by 2030, approximately 600 million people will still face hunger, a figure similar to that at the launch of the UN's 2030 Agenda in 2015 (FAO et al. 2023). Additionally, evidence suggests that the growth rate of agricultural productivity is slowing, and high pollution and emissions have pushed production capacity to its limits, resulting in land and environmental degradation (FAO 2021). The Food and Agriculture Organization (FAO) of the United Nations predicts that to meet global food demand by 2050, crop and livestock production must increase by 60% compared to 2006, with the majority of this growth coming from yield improvements.

In this context, the emphasis on improving agricultural productivity and maintaining ecosystem balance, as highlighted in SDG 2.4, becomes particularly crucial. Traditional agricultural production data often aggregate at the administrative unit level and lack detailed spatial distribution information, thereby limiting the precision of food production assessments. Big Earth Data technology, particularly Earth observation technology, can provide spatial patterns, areas, and yield information for different crop types, thereby offering a scientific data foundation for balancing agricultural production efficiency and environmental conservation.

This chapter focuses on assessing the global food production growth potential, with a particular emphasis on Southeast Asia, a region highly sensitive to food security. By integrating multi-source data, this chapter identifies and estimates the yields of various crops, with a specific focus on monitoring rice, a key crop in the region. Timely monitoring of global food production through data integration and technological innovation can provide scientific evidence for governments to formulate effective food policies. This not only promotes the improvement of existing agricultural production models but also supports the scientific planning of future food security strategies, ensuring food security and advancing the realization of the UN's 2030 Agenda.

2.2 Main Contributions

This chapter presents three case studies focused on the monitoring and evaluation of SDG 2.4 indicators, thereby contributing to the global progress of SDG 2 through achievements in three areas: data products, methods and models, and decision support (Table 2.1).

H. Guo, *Big Earth Data in Support of the Sustainable Development Goals (2023)—The Belt and Road*, Sustainable Development Goals Series, https://doi.org/10.1007/978-981-95-3178-3_2

Table 2.1 Cases and their main contributions

Targets	Tiers	Cases	Contributions
SDG 2.4 By 2030, ensure sustainable food production systems and implement resilient agricultural practices that increase productivity and production, that help maintain ecosystems, that strengthen capacity for adaptation to climate change, extreme weather, drought, flooding and other disasters and that progressively improve land and soil quality	Tier III	Spatiotemporal differentiation patterns of potential for yield increases of global crops	Data product: A dataset of the differences in per-unit yields of major global crops for 2010, 2015, and 2020 Method and model: By integrating multi-source data with machine learning methods, this approach aims to estimate the potential for yield increases in major global crops Decision support: This work is instrumental in ensuring global agricultural production and food security, and it provides critical information for the formulation of sustainable agricultural development policies
	Tier II	Crop identification and production estimates for Southeast Asia in 2020	Data product: A spatial distribution dataset of the area and yield of multiple crop types in Southeast Asian countries for 2020 Method and model: Cross-entropy distribution probability model Decision support: This work provides a detailed spatial quantification of food production in Southeast Asia in 2020, offering important scientific evidence for the formulation of region-specific food security policies
	Tier III	Extraction and analysis of interannual rice cultivation areas in Southeast Asia	Data product: A dataset of annual rice cultivation area extraction for five Southeast Asian countries from 2019 to 2021 Method and model: A large-area tropical/subtropical SAR-based annual rice mapping model that utilizes temporal statistical features and the U-Net model Decision support: This work provides supporting data for the assessment of sustainable agriculture in Southeast Asia, provides a theoretical basis for the identification of vulnerable areas for stable grain production, and serves as a reference for food trade policies by management authorities

2.3 Case Study

2.3.1 Spatiotemporal Differentiation Patterns of Potential for Yield Increases of Global Crops

Target: SDG 2.4 By 2030, ensure sustainable food production systems and implement resilient agricultural practices that increase productivity and production, that help maintain ecosystems, that strengthen capacity for adaptation to climate change, extreme weather, drought, flooding and other disasters and that progressively improve land and soil quality.

- **Background**

Food, as a basic necessity for human survival, not only serves as the cornerstone for promoting economic development and maintaining social stability but also increasingly poses a significant threat to national security and regional stability (Leng and Hall 2019; de S. Nóia Júnior et al. 2022). In recent years, due to multiple factors such as climate change, the degradation of natural resources, and population growth, food security issues have become more pronounced, garnering significant attention (Ray et al. 2015). The FAO estimates that in order to meet the global food demand in 2050, annual crop and livestock production needs to increase by 60% from the levels in 2006. Among this required growth, approximately 80% must be achieved through improved yields. Therefore, improving crop yields is the most crucial approach to addressing future food security issues.

Yield gaps, defined as the difference between the potential yield that can be achieved under optimal management conditions and the actual yield, are an important method for evaluating the potential for yield increases (Evans and Fischer 1999; Liu et al. 2012). Existing studies have analyzed yield gaps at regional or global scales (Liu et al. 2012; Mueller et al. 2012; Zhang et al. 2014; Gerber et al. 2024). However, discrepancies in data and methodologies used make it difficult to compare yield gaps across different regions (Fischer 2015; Licker et al. 2010). In global yield gap studies, commonly used statistical models often oversimplify crop growth processes, increasing the uncertainty in estimating potential yield increases (Neumann et al. 2010; Mueller et al. 2012; Gerber et al. 2024). These challenges limit the ability of previous research to provide sufficient scientific evidence for assessing the potential for yield increases under the framework of the UN SDGs. Therefore, it is necessary to develop a feasible method using Big Earth Data technology to measure the yield gaps for major global crops (namely wheat, rice, maize, and soybean) and to explore the spatial potential for yield increases for these four crops worldwide. Improving the accuracy and scientific validity of the assessment results will provide more reliable information for policymakers to formulate effective food security strategies and address future challenges.

- **Data**
  - Crop yield data: Crop yield data from 1961 to 2021 for countries worldwide, provided by the FAO.
  - Multi-scenario crop yield data: Data from 1980 to 2012, provided by the Agricultural Model Intercomparison and Improvement Project (AgMIP) and the Inter-Sectoral Impact Model Intercomparison Project (ISIMIP), with a spatial resolution of 0.5°.
  - Meteorological data: Global meteorological elements (such as precipitation, radiation, etc.) data from 1901 to 2021, provided by the Climatic Research Unit (CRU) at the University of East Anglia.
  - Irrigation, fertilization, and population data: Data regarding irrigation, fertilization, and population from 1961 to 2021 for countries worldwide, provided by the FAO.

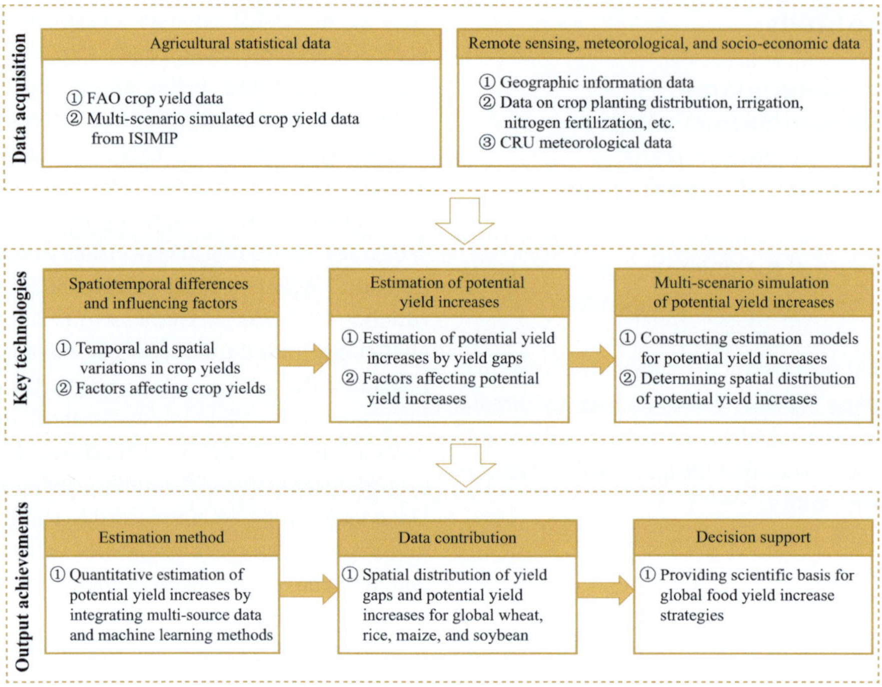

Fig. 2.1 The technical route for this case study

- **Method**

The technical route for this case study is illustrated in Fig. 2.1. Crop yield data for the four major crops from 1961 to 2021 were collected and organized, allowing for the analysis of yield variation characteristics of different crops in different countries during this period. Furthermore, using machine learning methods, we identified the key influencing factors of crop yield changes from 1961 to 2021, considering natural, economic, and social factors.

AgMIP and ISIMIP provided multi-scenario crop yield data by global gridded crop models (GGCMs). Simulations were conducted under varied harmonization and irrigation scenarios; for instance, irrigation scenarios included rainfed and full irrigation, and nitrogen scenarios included actual nitrogen application and sufficient nitrogen application. Based on these data, we constructed four combination scenarios of yield data: "rainfed + actual nitrogen application", "rainfed + sufficient nitrogen application", "sufficient irrigation + actual nitrogen application", and "sufficient irrigation + sufficient nitrogen application". Subsequently, we utilized the spatial production allocation model (SPAM) to obtain the areas of rainfed and irrigated cropland, and weighted the gridded yield data under these four scenarios to the national scale. This allowed us to derive national scale yield data under baseline, sufficient irrigation, sufficient nitrogen application, and sufficient irrigation with sufficient nitrogen application scenarios. Finally, using the yield gap method, we calculated the potential yield increases for different crops in various countries worldwide, including potential yield increases under sufficient irrigation, potential yield increases under sufficient nitrogen application, and overall potential yield increases. This analysis enabled the determination of the global spatial potential for yield increases across different crops.

- **Results and Analysis**

1. **Global Major Crop Yield Status**

In this case study, by comparing the yields and harvested areas of four major crops from 1961 to 2021 with the data from 1961, we analyzed the variation characteristics of crop yields and harvested areas over these 60 years. From 1961 to 2021, the yields of major crops worldwide have grown rapidly. Compared to 1961, the yields of maize and wheat in 2021 had more than 2 times, while the yields of rice and soybean had more than 1 time as well. Regarding harvested areas, the increase in the soybean area is particularly notable, with the 2021 harvested area exceeding 4 times that of 1961. Meanwhile, the harvested areas for maize, wheat, and rice have shown relatively minor changes, with increases of less than one-fold (Fig. 2.2).

The yield increments of each crop from 1961 to 2021 exhibit significant spatial heterogeneity. In China, the yields of all four major crops have significantly increased, with the yields of maize and wheat in 2021 being more than 3 times higher than those in 1961, and the yields of rice and soybean have increased more than 2 times. In most countries in Southeast Asia, Western Asia,

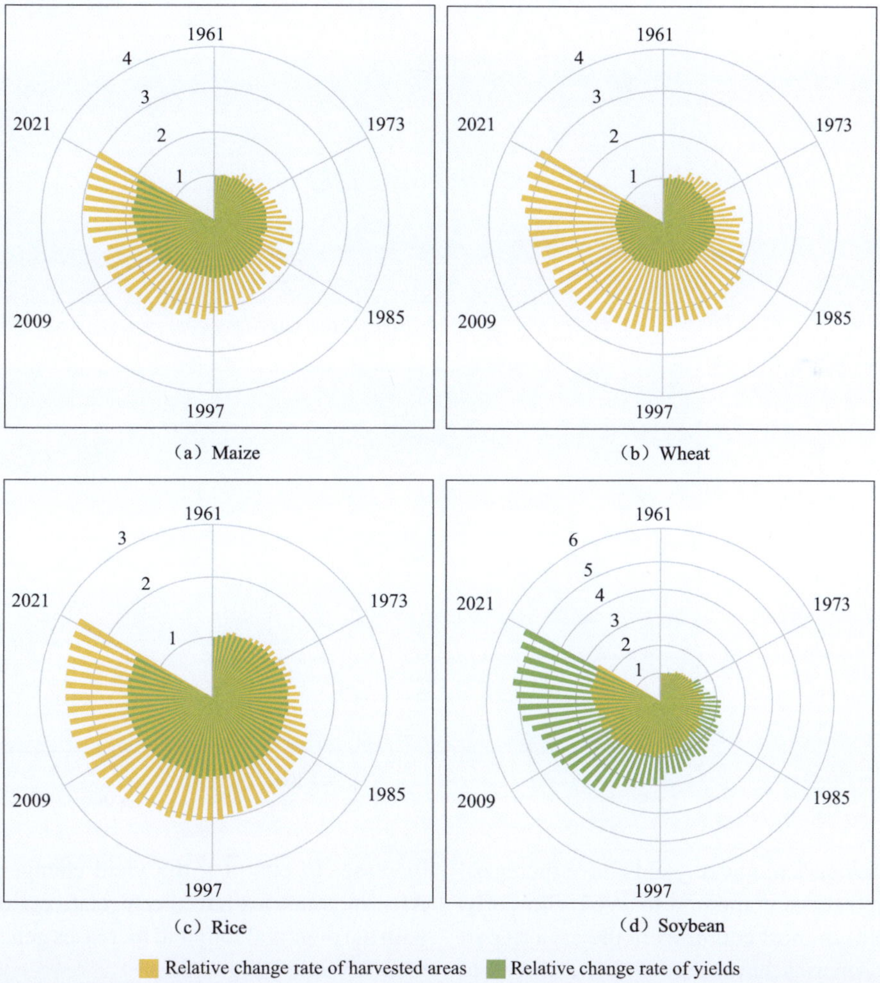

Fig. 2.2 Changes in global major crop yields and harvested areas from 1961 to 2021. *Note* Relative change rate = data in specific year/data in 1961

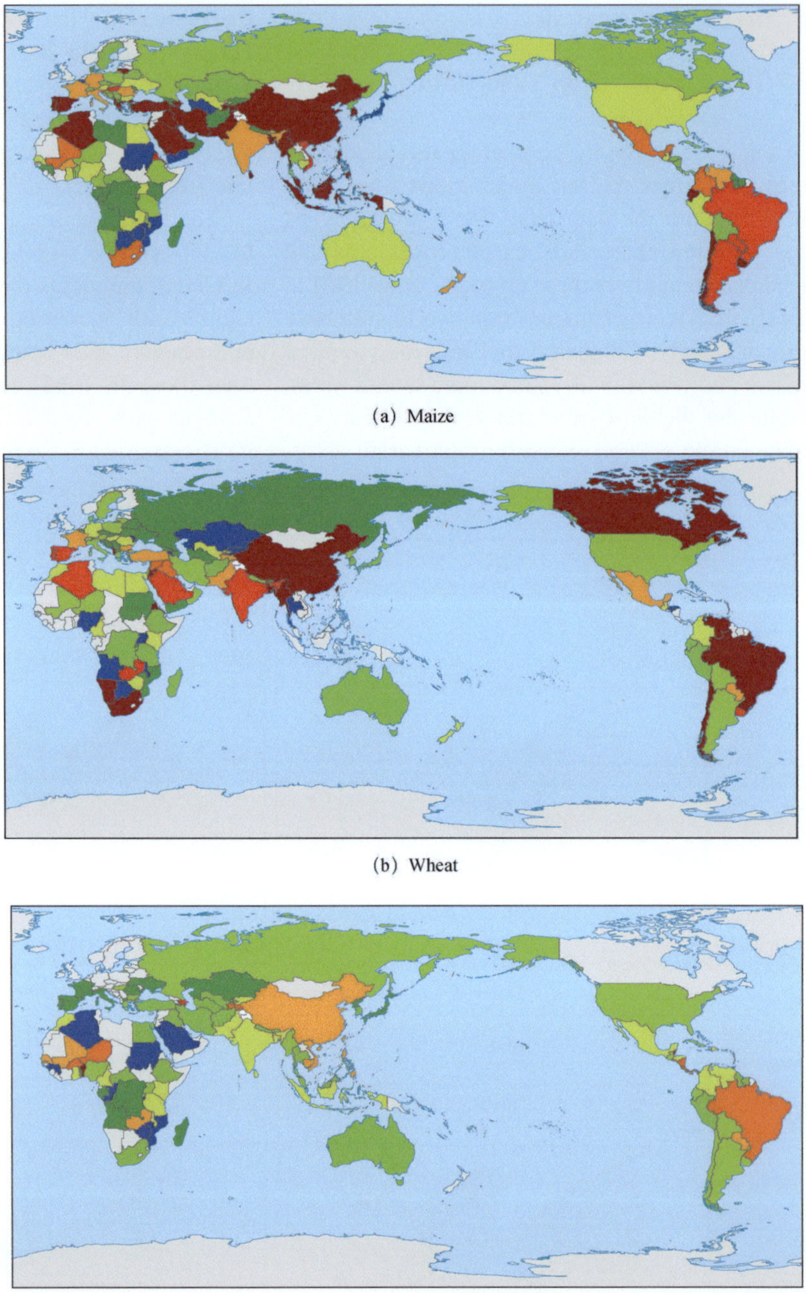

(a) Maize

(b) Wheat

(c) Rice

Fig. 2.3 The spatial distribution of changes in the yields of the main global crops in 2021 compared to 1961. *Note* Relative change rate = single crop yield data in 2021/data in 1961

and South America, maize yields have increased more than 3 times compared to 1961. Similarly, wheat yields in most countries in the eastern part of South America have also shown a threefold increase. In contrast, the yield changes in most African countries have been relatively moderate, with the potential for yield increases generally less than 1.5 times (Fig. 2.3).

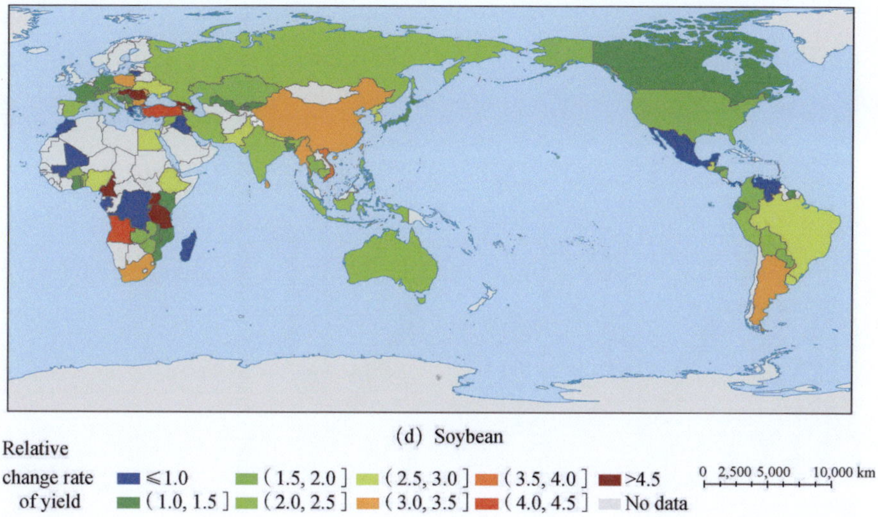

Fig. 2.3 (continued)

2. **Influencing Factors of Global Major Crop Yields**

This case study utilized factors such as temperature indices (maximum, minimum and average temperatures, frost days, and minimum frost days), water indices (precipitation, vapor pressure, and potential evapotranspiration), radiation index (cloud cover), management practices (irrigated area and nitrogen application rate), and population, based on the random forest (RF) model to simulate the yields of major crops in various countries around the world in 2020. The simulated results were then compared with actual statistical yields (as shown in Fig. 2.4). The results demonstrate that the RF model had high simulation accuracy for crop yields in 2020, with R^2 values exceeding 0.8, and the root mean square error (RMSE) and mean absolute error (MAE) were significantly lower than the average crop yields. This indicates that the RF model, which integrates factors such as growing season climate factors, statistical data, and management measures, is highly applicable for simulating crop yields and can provide high-precision simulation results.

Figure 2.5 presents the relative importance ranks of the variables in the RF models for the four crops. Temperature indices, water indices, and radiation index are combined from 18 climate factors. Among the management measures, the nitrogen application rate has a relatively higher importance for all four crops, while the irrigated area has a relatively lower importance. Among the climate factors, water indices and temperature indices are the primary factors influencing the yields of the four crops. Specifically, water indices have the most significant impact on the yields of maize, rice, and soybean, while temperature indices have the strongest influence on wheat yields.

Figure 2.6 illustrates the differences in key driving factors for the five major producing countries of the four major crops. Among the five major producing countries for each crop, the irrigated area, nitrogen application rate, and population are primary influencing factors on crop yields. For example, the irrigated area is most important for maize, wheat, and rice in China, maize and wheat in India, and wheat production in France. This contrasts with the primary driving factors of global crop yields, where overall yields are still predominantly influenced by climate factors (Figs. 2.5 and 2.6). Regarding climate factors, the water indices have substantial impacts not only on crop yields in South America (such as maize in Brazil, and soybeans in Brazil, Paraguay, and Argentina), but also on wheat production in Russia and France.

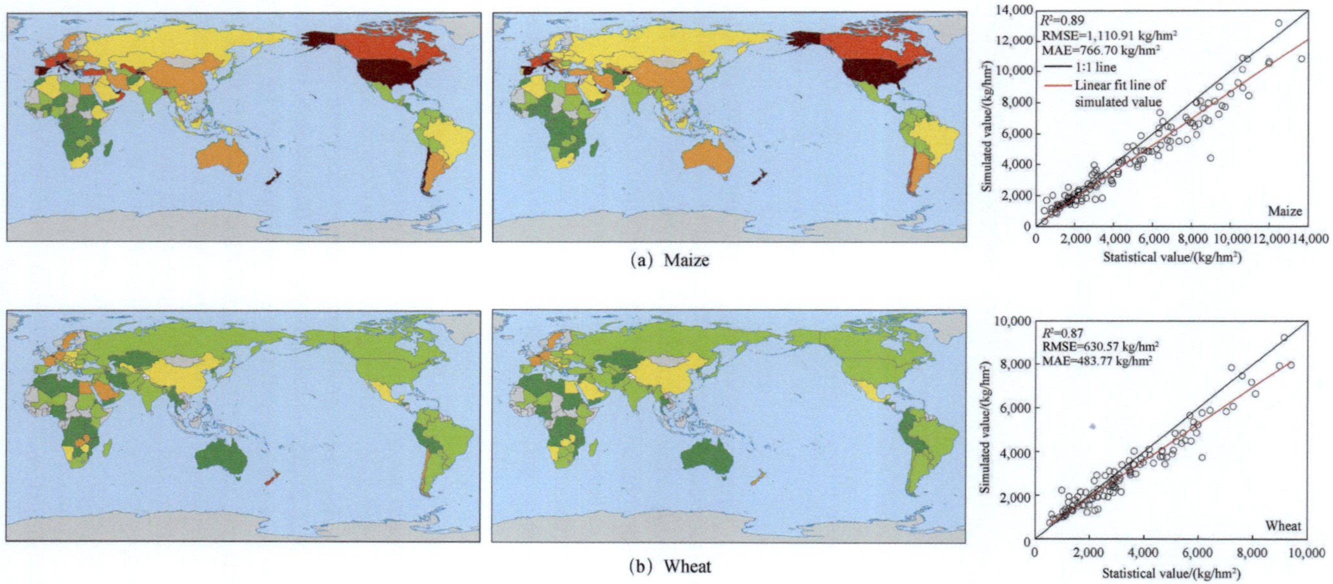

Fig. 2.4 Global comparison of simulated values and statistical values of global major crop yields. *Note* The left-hand panel shows the statistical yield, and the right-hand panel shows the simulated yield

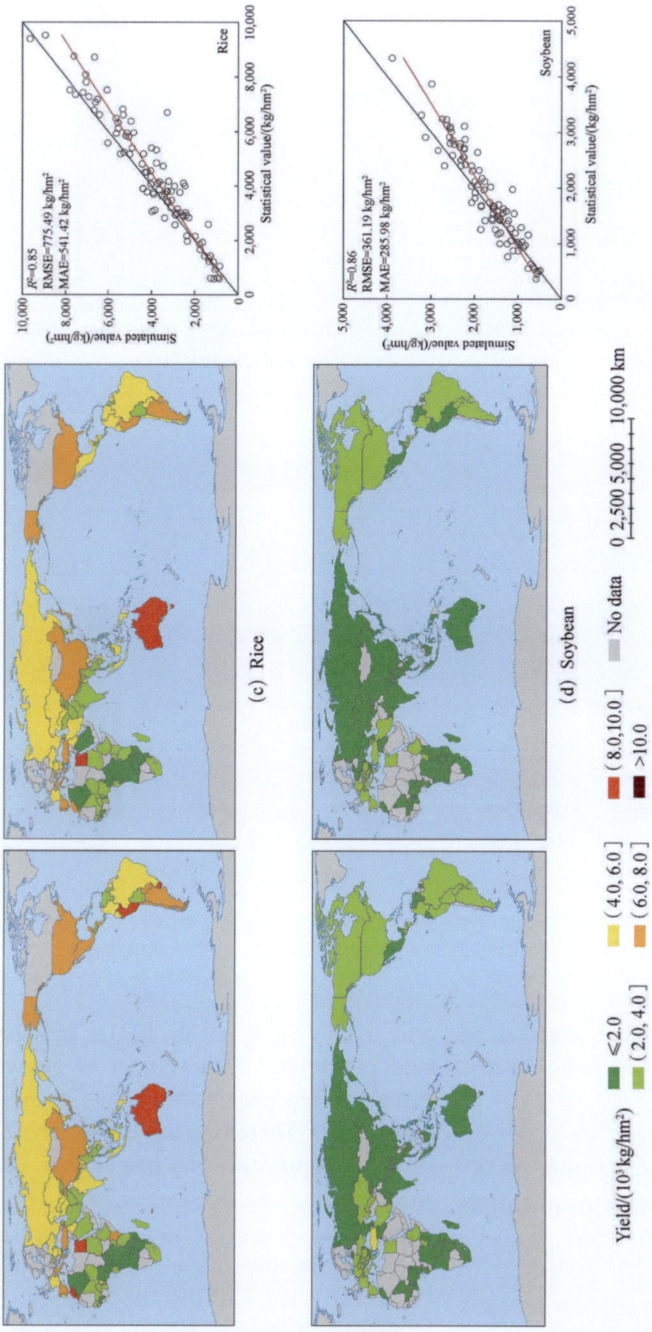

Fig. 2.4 (continued)

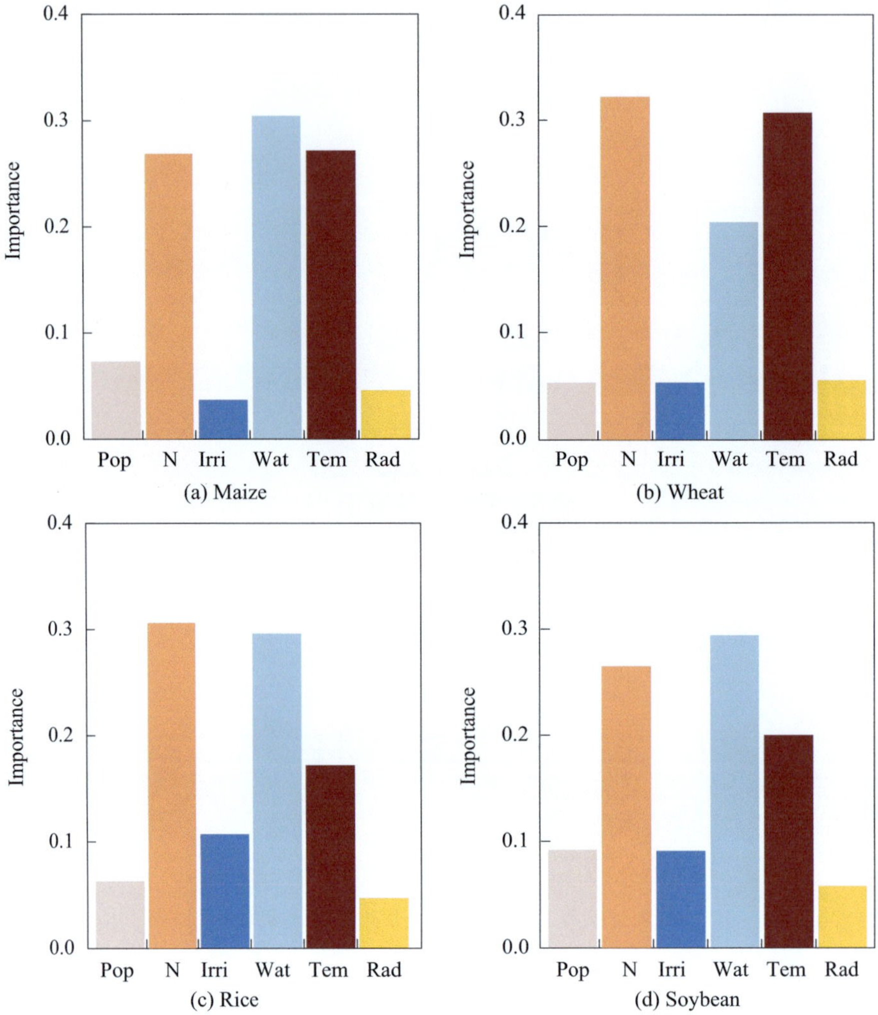

Fig. 2.5 Relative importance of each driving factor of the RF model. *Note* Pop: population, N: nitrogen application rate, Irri: irrigated area, Wat: water indices, Tem: temperature indices, and Rad: radiation indice

The temperature indices also affect crop yields, particularly for soybean production in the United States and Paraguay (Fig. 2.6).

3. **Potential for Yield Increases Under Sufficient Irrigation and Nitrogen Application Scenarios**

Variety improvement, nitrogen application, and irrigation are the three primary measures for increasing yields (Zabel et al. 2021; Mueller et al. 2012). However, breeding varieties and their widespread adoption require extensive time and trials (Xiong et al. 2021), and the current rate of variety improvement is already struggling to keep pace with climate change (Zhang et al. 2022). Therefore, this case study focuses on nitrogen application and irrigation as convenient methods that can achieve yield increases in the short term.

Based on the output results from GGCMs of AgMIP and ISIMIP, the potential for yield increases of maize, wheat, rice, and soybean under sufficient irrigation or nitrogen application scenarios was determined for different countries and regions. The results indicate significant spatial

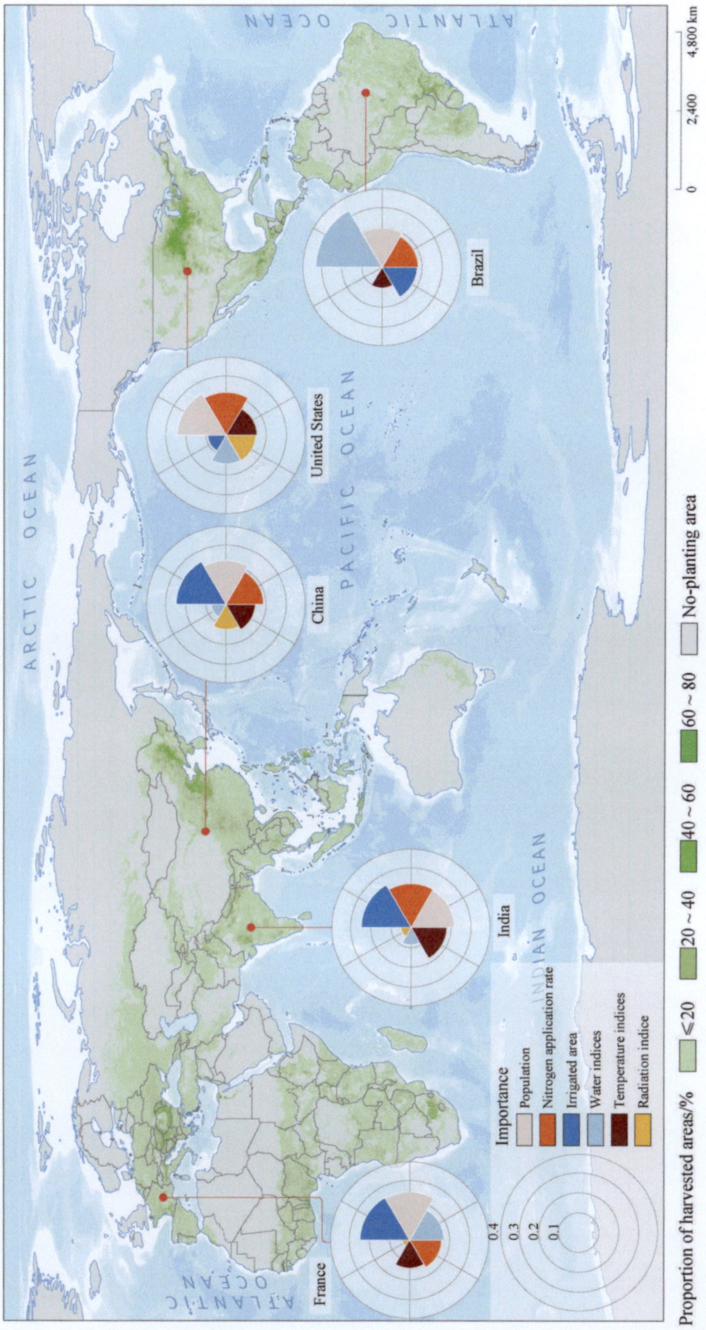

(a) Maize

Fig. 2.6 Importance ranking of yield drivers in the five major producing countries of each crop

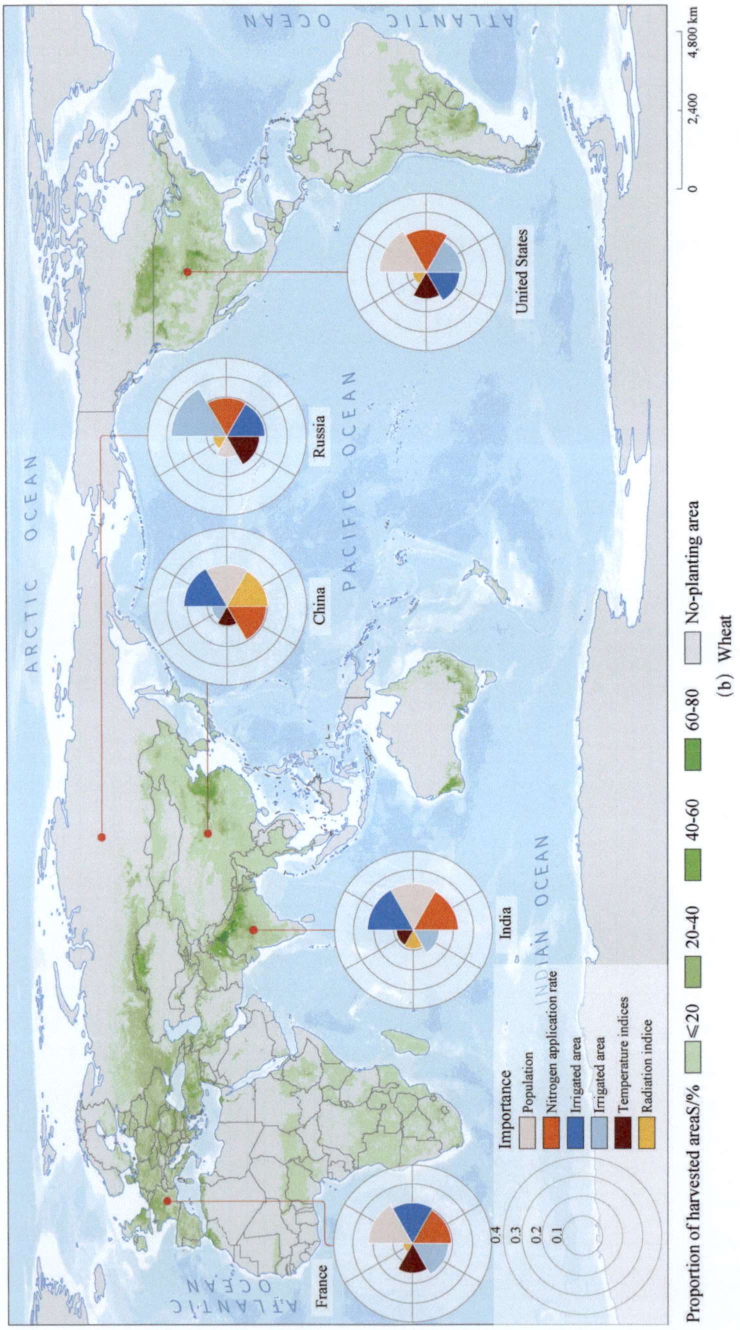

Fig. 2.6 (continued)

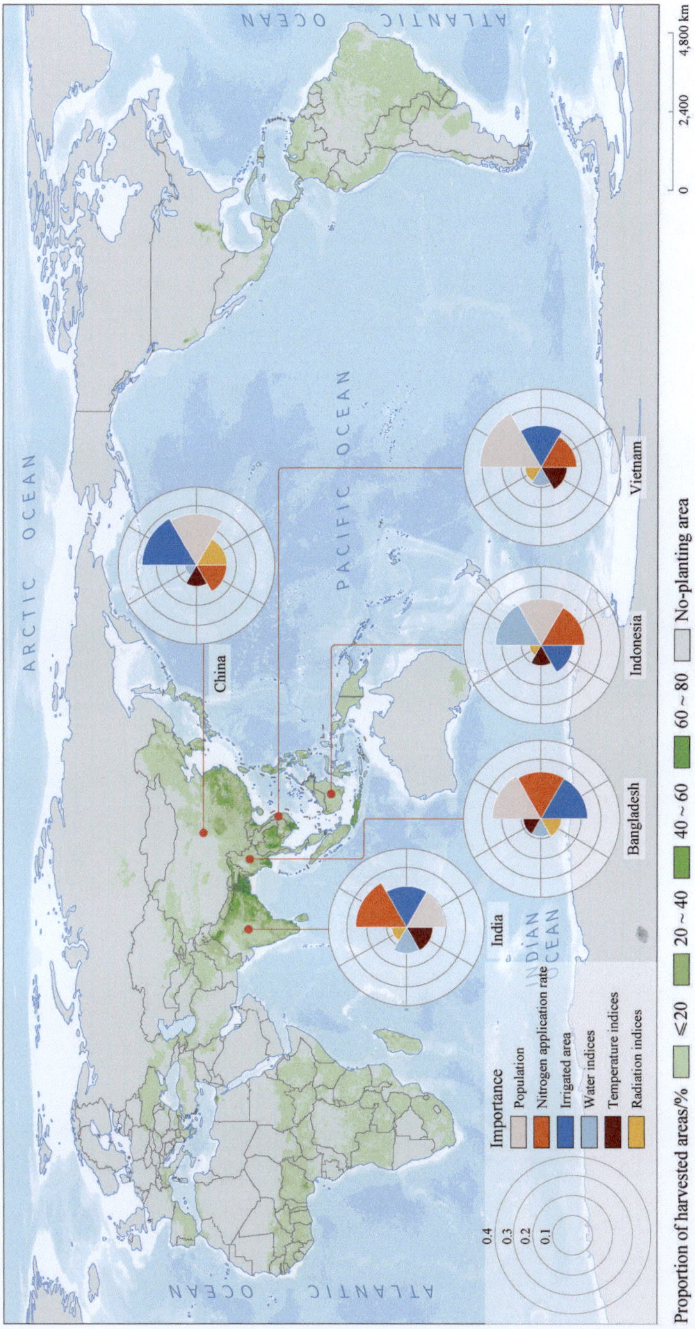

(c) Rice

Fig. 2.6 (continued)

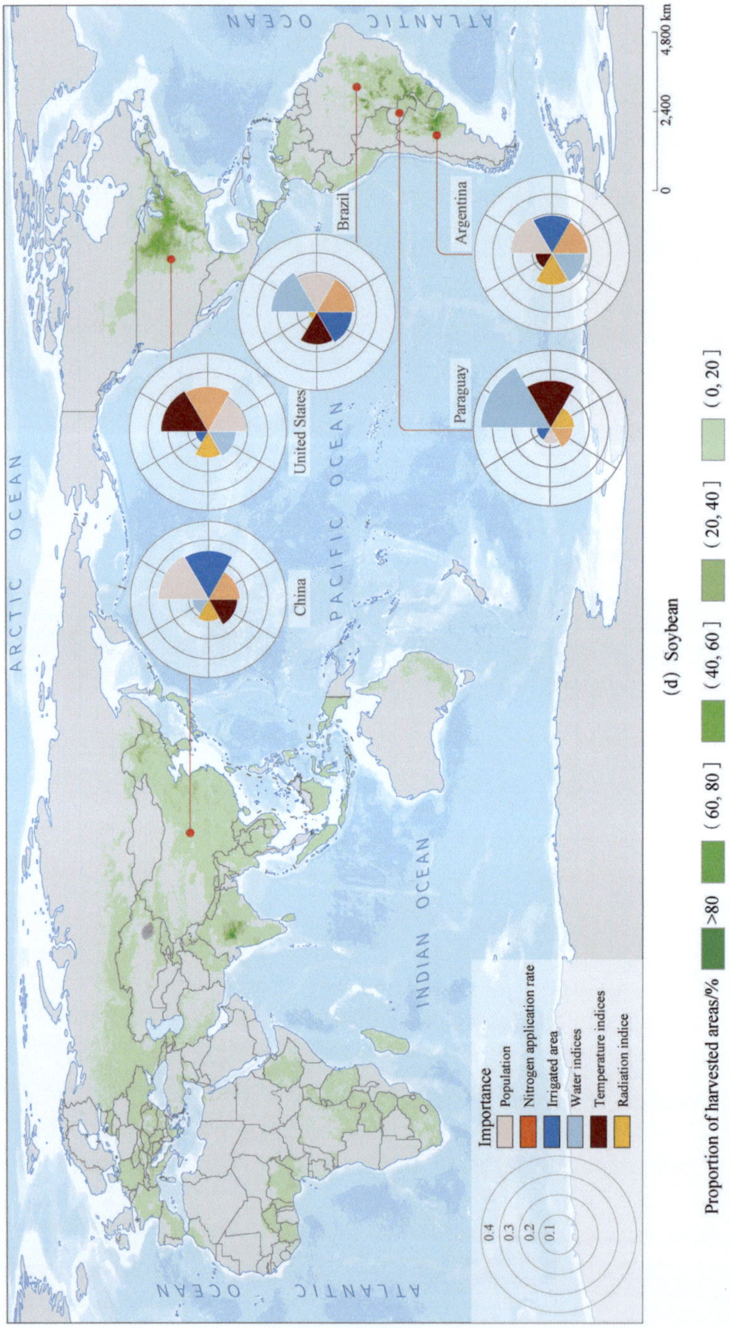

Fig. 2.6 (continued)

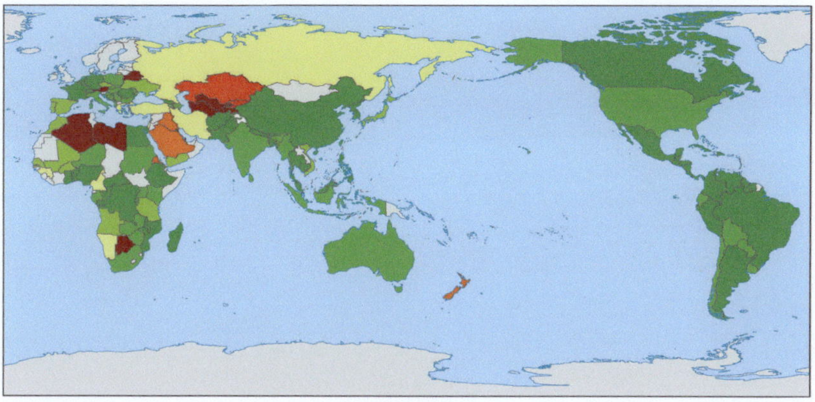

(a) Maize

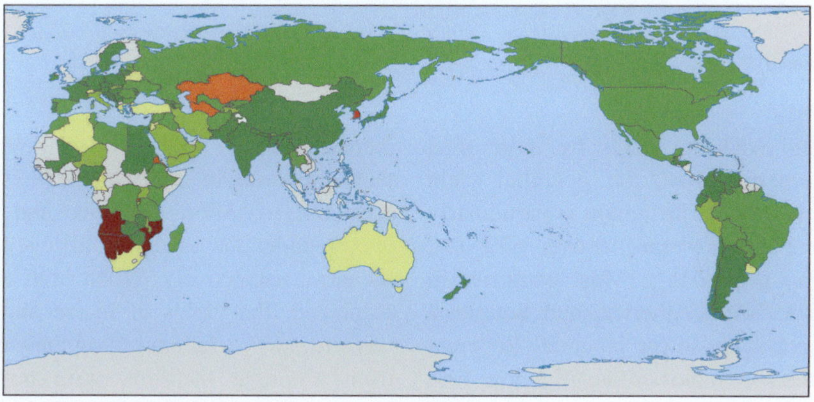

(b) Wheat

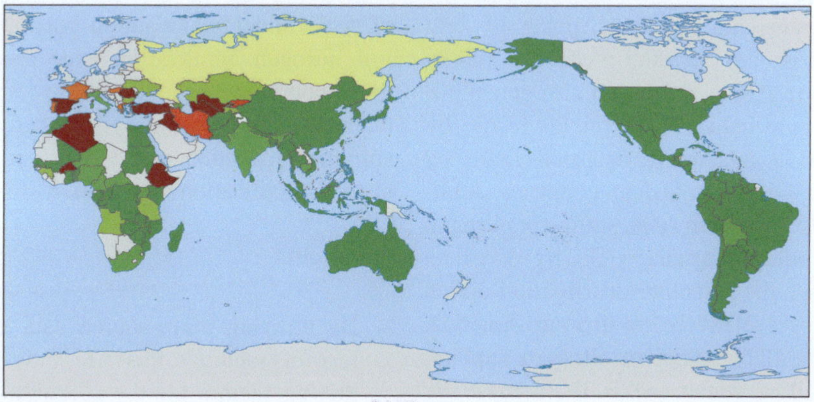

(c) Rice

Fig. 2.7 Spatial distribution map of potential yield increases of four major crops under the sufficient irrigation scenario

differences in the potential for yield increases of various crops (Figs. 2.7 and 2.8).

Under the sufficient irrigation scenario, regions with substantial potential for yield increases for the four major crops are mainly concentrated in Northwestern Asia, Eastern Europe, and Southern Africa. In most countries of Eastern Europe and Northwestern Asia, maize yields could increase by more than 40%, wheat yields by 20% to 60%, rice yields by more

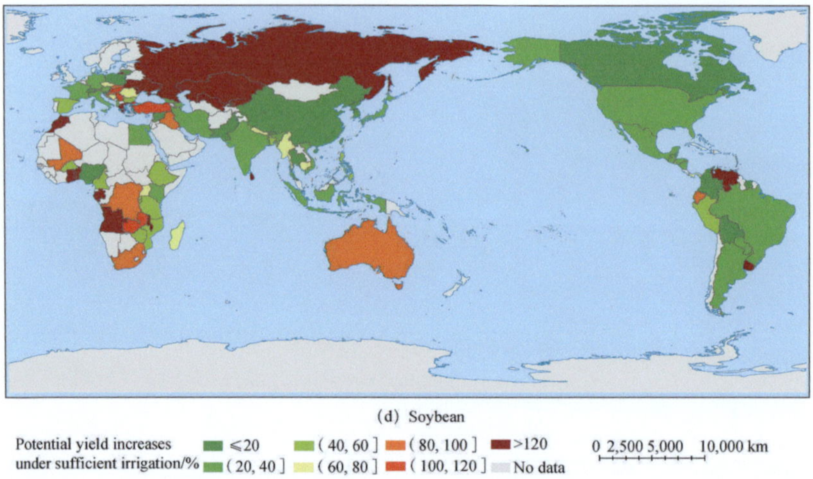

Fig. 2.7 (continued)

than 40%, and soybean yields by more than 60%. In Southern Africa, the potential yield increase under sufficient irrigation is particularly high for wheat and soybean, at over 60% and 40% to 100%, respectively. Most countries in South America, North America, and Southeast Asia have a relatively lower potential for yield increases of the four major crops, all below 40% (Fig. 2.7).

Under the sufficient nitrogen application scenario, regions with substantial potential yield increases for maize, rice, and wheat are mainly concentrated in Africa. Increased nitrogen application rate could induce wheat yields to rise by more than 61%, and maize and rice yields by more than 41% in most African countries. Additionally, in Northwestern Asia, sufficient nitrogen application could boost maize yields by more than 41%. Although the potential for irrigation-driven yield increases is relatively low in South America, mostly below 40%, sufficient nitrogen application could allow wheat yields to increase by 41% to 80% in most countries in South America (Fig. 2.8).

Globally, the average potential yield increases of the four major crops under the sufficient irrigation scenario range from 13.7% to 29.8%, and under the sufficient nitrogen application scenario, from 2.9% to 39.1%. Maize and wheat demonstrate significant potential for yield increases under both scenarios, with potential yield increases of 21.1% and 28.7% under the sufficient irrigation scenario, and 24.0% and 39.1% under the sufficient nitrogen application scenario, respectively. When both measures are combined, the yields of maize and wheat will further improve, far exceeding the yield increases from a single measure, with potential yield increases reaching 59.0% and 97.8%, respectively. Soybean has a potential for yield increases of 29.8% under full irrigation, but only 2.9% under full nitrogen application. Rice demonstrates a lower potential for yield increase compared to the other three crops, with potentials of 13.7% under sufficient irrigation and 17.2% under sufficient nitrogen application (Fig. 2.9).

Highlights

- By utilizing multi-source data and machine learning methods, this case study identified the driving factors for the variation in yields of global major crops and produced datasets of yield gaps for these crops for the years 2010, 2015, and 2020. These datasets provide crucial data support for the assessment of SDG 2.4 at the national scale.
- There is still significant potential for yield increases in major global crops. Appropriate nitrogen fertilizer application in rice

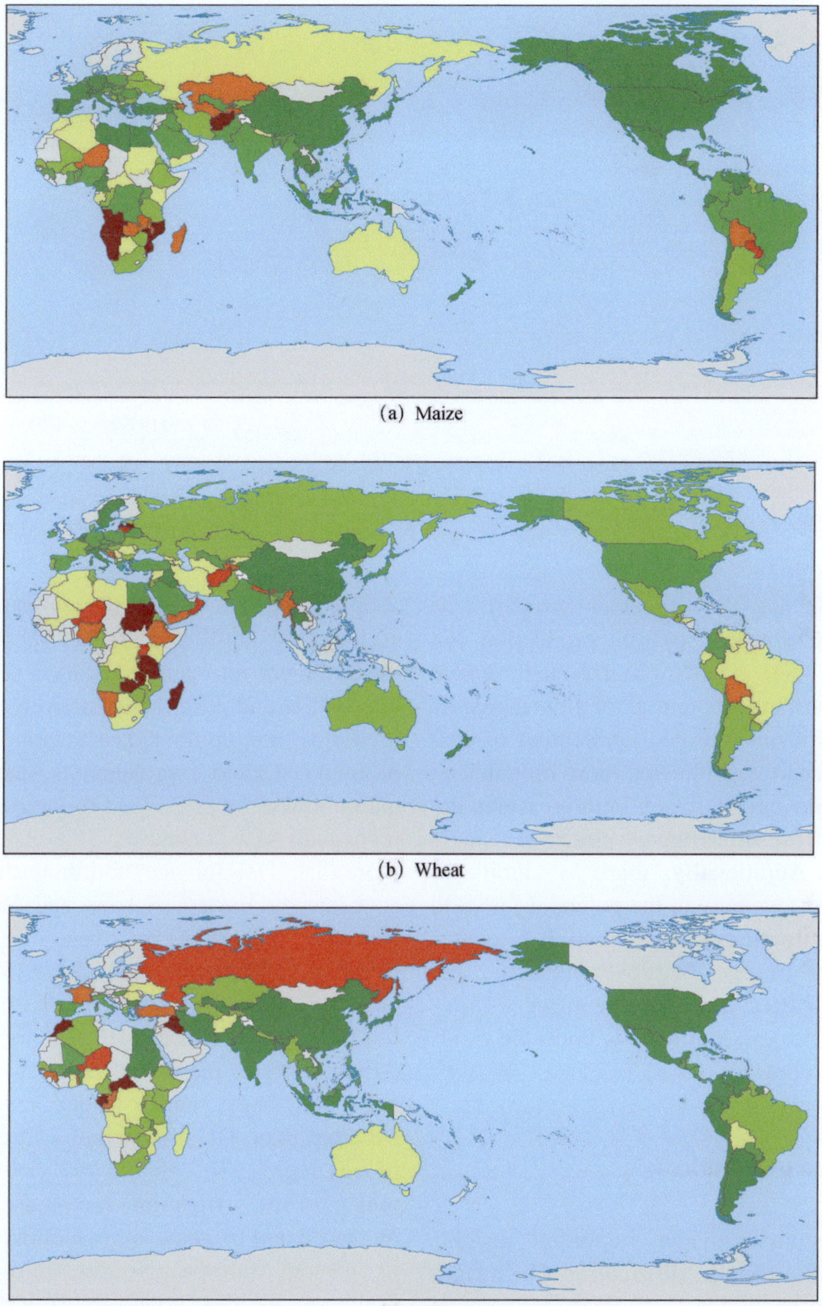

(a) Maize

(b) Wheat

(c) Rice

Fig. 2.8 Spatial distribution map of the potential for yield increases of four major crops under the sufficient nitrogen application scenario

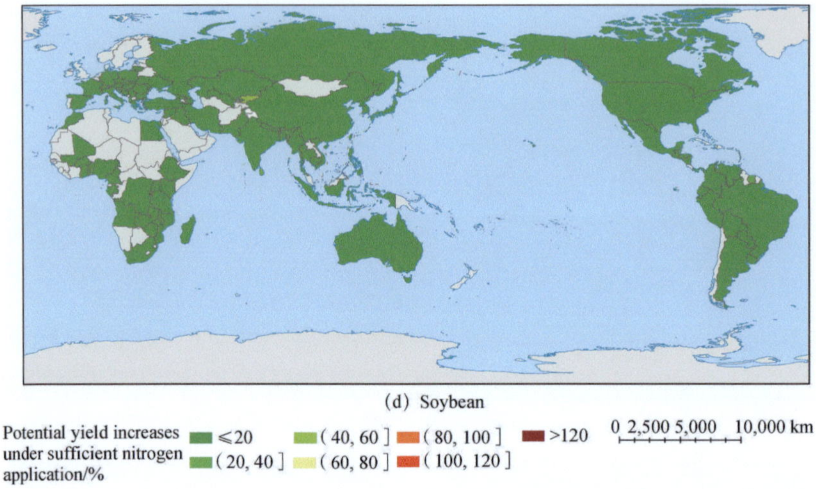

Fig. 2.8 (continued)

production and the implementation of irrigation measures in soybean production can enhance their yields. Maize and wheat exhibit substantial potential for yield increases under sufficient nitrogen application or irrigation scenarios, with even more pronounced yield improvements when both measures are combined, far surpassing the effects of a single measure. Additionally, there is significant spatial heterogeneity in the potential for yield increases. Implementing irrigation measures in Eastern Europe, Northwestern Asia, and Southern Africa, and applying nitrogen fertilizer in Africa can significantly boost the yields of major crops.

- **Discussion and Outlook**

This case study provides important insights into the spatiotemporal distribution of changes in global crop yields and their influencing factors. This is of significant importance for formulating agricultural development policies, promoting sustainable agricultural development, and achieving food security goals. The study indicates that major crops still have considerable potential for yield increases. Specifically, maize and wheat could increase by 21.1% and 28.7% respectively under the sufficient irrigation scenario; by 24.0% and 39.1% respectively under the sufficient nitrogen application scenario; and by 59.0% and 97.8% respectively when both scenarios are combined, showing even more pronounced yield improvements. Increasing irrigation measures could lead to yield increases of more than 41% for maize, 21%-60% for wheat, more than 41% for rice, and more than 61% for soybeans in Eastern Europe and Northwestern Asia. In Southern Africa, these measures could result in yield increases of over 61% for wheat and 41%-100% for soybean. Nitrogen limitation is mainly concentrated in most parts of Africa, where increased nitrogen application rate could boost maize, wheat, and rice yields by more than 41%, 61%, and 41% respectively in this region. To enhance the yields of global major crops, irrigation measures should be increased and irrigated areas should be expanded in Eastern Europe, Northwestern Asia, and Southern African countries. Additionally, an appropriate nitrogen application rate should be implemented in the production of maize, wheat, and rice in most African countries. Future efforts should focus on strengthening scientific conservation and management practices to develop more effective strategies for improving global crop yields.

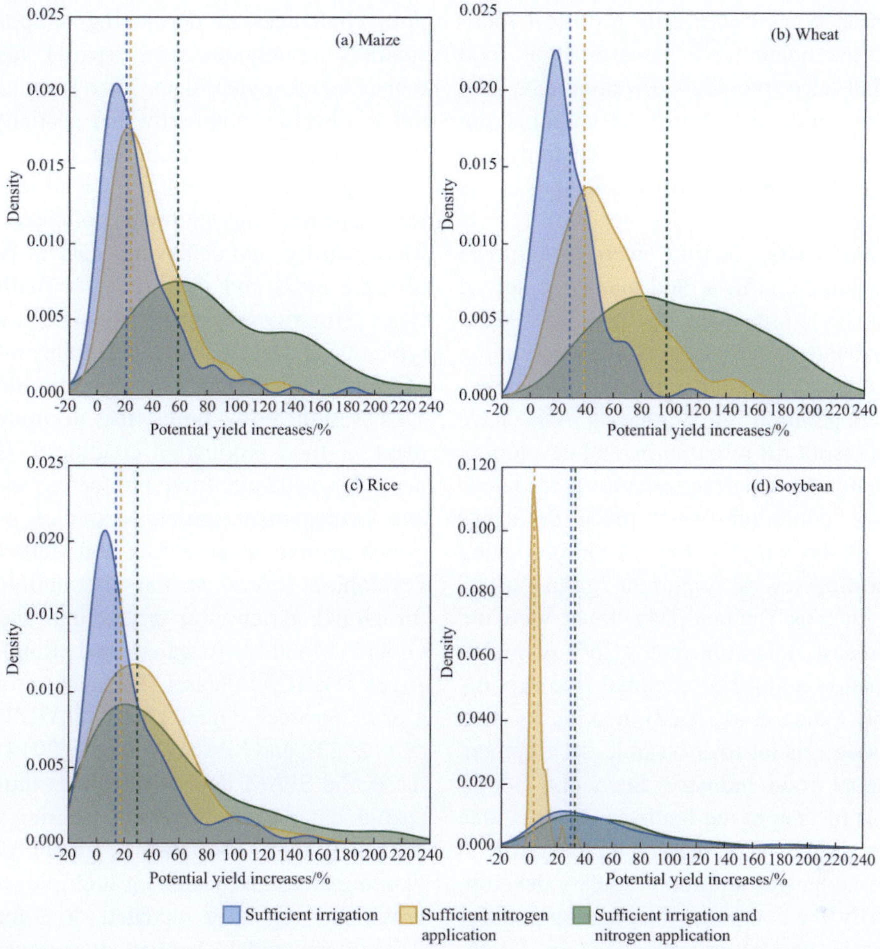

Fig. 2.9 Distribution of the average potential for yield increases of the four major global crops. *Note* Yield gap represents the difference between actual farm yield and potential yield achievable under optimal management conditions. The blue, yellow and green dashed lines denote the global average potential increase under full irrigation, full nitrogen application or both, respectively

2.3.2 Crop Identification and Production Estimates for Southeast Asia in 2020

Target: SDG 2.4 By 2030, ensure sustainable food production systems and implement resilient agricultural practices that increase productivity and production, that help maintain ecosystems, that strengthen capacity for adaptation to climate change, extreme weather, drought, flooding and other disasters and that progressively improve land and soil quality.

- **Background**

"Zero hunger" is one of the key goals of the SDGs, and SDG 2.4 further emphasizes the need to enhance productivity and yields while maintaining ecosystems and achieving sustainable food production (Lipper et al. 2020). Changes in the planting areas of different crop types and a series of human activities related to increasing food production can significantly impact landscape patterns, structures, and functions of ecosystems. Therefore, timely and accurate acquisition of planting areas and yields for different crop types is not only a critical foundation for the quantitative assessment of food security, but also provides essential data and scientific evidence to support decision-making aimed at balancing agricultural productivity and environmental protection (Becker-Reshef et al. 2023).

In recent years, factors such as climate change, regional conflicts, and pandemic spread have seriously threatened global food security. The abundant agricultural products, superior geographical locations, and enhanced international cooperation in Southeast Asia have made it a hotspot for researching and developing SDG 2. Firstly, the superior agricultural resources and climate conditions have made Southeast Asia one of the world's major food-producing regions, hosting several important "grain basket" countries (such as Thailand, Myanmar, Vietnam, etc.). Southeast Asia contributes 26% of global rice production and 40% of global rice exports, respectively (Yuan et al. 2022), making its food security issue crucial to the stable development of the global food industry. Secondly, Southeast Asia is located at the land-sea junction area covered by China's BRI, playing a crucial role in food trade exchanges and food industry development. Finally, the complex crop planting patterns and intensive agricultural practices in Southeast Asia also impact many critical ecological indicators, such as methane emissions and soil quality in cropland. Therefore, the stability of grain planting areas and yields in Southeast Asia is of great significance for maintaining global food trade, ensuring global food security, and protecting the sustainable development of the environment.

However, it is difficult to improve the current cropland area, cropland quality, and grain production in Southeast Asia. In particular, the occupation and loss of cropland resources, along with the scarcity of irrigation resources, pose significant challenges to ensuring food security in this region (D'Amour et al. 2017). Moreover, the extensive crop planting area, complex cropping structure, and significant geographical heterogeneity in Southeast Asia pose challenges in developing temporally and spatially continuous crop spatial distribution maps and crop yield data. Therefore, it is essential to develop temporally and spatially continuous crop spatial distribution maps and crop yield products, providing scientific evidence for adjusting agricultural policies, ensuring food security, and achieving SDG 2. Traditional planting areas and yield data for multiple crop types primarily rely on statistical data, which are typically aggregated at the administrative unit, lacking detailed spatial distribution information. This significantly limits the accurate assessment of food production conditions. Given the need for accurate food production assessment and management, previous studies have integrated remote sensing data and statistical data, developing several spatial distribution datasets for global agricultural production, such as the Global Monthly Irrigated and Rainfed Crop Areas (MIRCA) dataset (Portmann et al. 2010), global agro-ecological zones (GAEZ) (Grogan et al. 2022), and SPAM (You et al. 2014). Among these, the SPAM dataset integrates multi-source spatial data, such as remote sensing data and statistical data, to generate global pixel-level planting areas and yields for multiple crop types, providing regularly updated long-term data. It has been widely applied in various research fields globally (Dadrasi et al. 2023). Therefore, it is a crucial method to effectively employ SPAM data for accurate monitoring and managing food production at a large scale, further ensuring food security and advancing the achievement of SDG 2.

This case study analyzed the spatial distribution of planting areas and yields for multiple crop types in Southeast Asia based on the 2020 SPAM dataset and conducted a spatiotemporal analysis of food production changes by integrating the 2010 SPAM dataset. This enables multidimensional monitoring and evaluation of food production, providing scientific data and decision-making support for optimizing agricultural resource allocation and promoting sustainable agricultural production in Southeast Asia.

- **Data**
  - The 2020 and 2010 SPAM datasets: Spatial distribution of harvested areas and total production for more than 40 crop types with a spatial resolution of 10 km, provided by the International Food Policy Research Institute (IFPRI).
  - The 2020 and 2010 Gridded Population of the World data for Southeast Asia, with a spatial resolution of 1 km, provided by the National Aeronautics and Space Administration (NASA)'s Socioeconomic Data and Applications Center.

- **Method**

In this case study, ten Southeast Asian countries were used as the study area (Fig. 2.10), and the SPAM datasets for 2010 and 2020 were used, supplemented by spatially gridded data on population. Spatiotemporal characterization and comprehensive assessment of the areas and yields of all crops (i.e., wheat, rice, maize, barley, millet, sorghum, pearl millet and other cereals) and major crops (i.e., wheat, rice, and maize) in Southeast Asia were conducted. SPAM datasets were generated by integrating multiple sources of spatial information data such as cropland distribution, area and yield statistics of crop types, agricultural irrigation distribution, and crop suitability distribution. By utilizing the principle of cross-entropy, the areas and yields of crops were precisely assigned to the pixel scale, so as to simulate the spatiotemporal distribution of crops at the regional and even global scales, which has the advantages of high accuracy and spatiotemporal continuity (Yu et al. 2020). In this case study, we first selected all crop types from the SPAM 2020 dataset for spatial distribution analysis of harvested area and total production of all crops and major crops, and combined it with the standard deviation ellipse method for spatial distribution of the center of gravity of major crops. Second, the spatial distribution characteristics of total crop harvested area and total production per capita were further analyzed in conjunction with spatially gridded population data that were aggregated to 10 km spatial resolution. Finally, spatiotemporal changes in total production and total production per capita of all crops were analyzed based on SPAM 2010 and 2020 datasets. In order to describe the changes in food production more clearly, this case quantified the changes in food production into 3 levels based on the overall distribution of data and previous studies (Chen et al. 2011; Li et al. 2014). Annual variability $\leq -1\%$ was defined as a decreasing trend, -1% to 1% as a stable trend, and $\geq 1\%$ as an increasing trend at the pixel scale.

- **Results and Analysis**

1. **Harvested Areas and Total Production of Crops and Major Crops in the Ten Southeast Asian Countries in 2020**

The distribution of harvested areas and total production of crops in the ten Southeast Asian countries in 2020 is shown in Fig. 2.11. In the central and southern regions, total harvested areas of crops at the pixel scale were primarily below 2,000 hm^2 (Fig. 2.11a), whereas the northwestern region contained more pixels with harvested areas exceeding 2,000 hm^2. Rice is the crop with the largest harvested area in Southeast Asia (Fig. 2.11d), followed by maize (Fig. 2.11e) and wheat (Fig. 2.11c). As the global key rice-exporting region, the pixel-level rice harvested areas in most regions in the northwestern part of Southeast Asia exceeded 3,000 hm^2. Maize harvested area was concentrated in specific regions of Southeast Asia, such as Southwestern Indonesia and Southern Philippines, with pixel-level maize harvested areas surpassing 700 hm^2. In 2020, the total production of crops at the pixel scale in Southeast Asia was primarily below 5,000 t (Fig. 2.11b), and higher total production areas were mainly located in the northwestern part of Southeast Asia. Among the three major

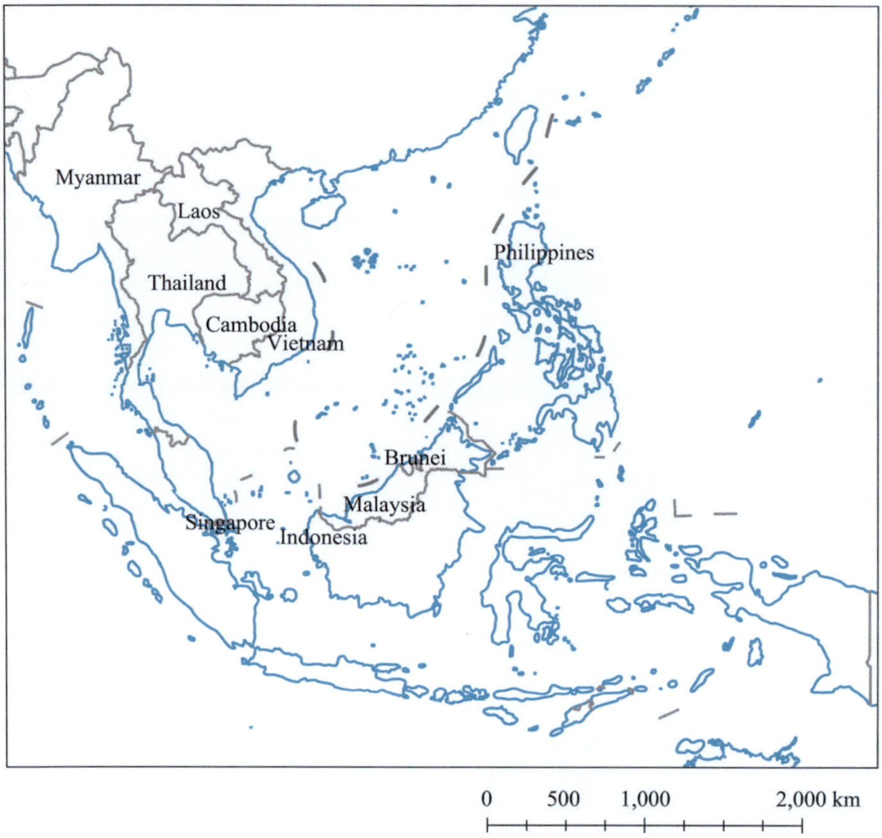

Fig. 2.10 The study area for this case study: ten Southeast Asian countries

crops, the total production of rice (Fig. 2.11g) and maize (Fig. 2.11h) were significantly higher than that of wheat (Fig. 2.11f).

To characterize the spatial directionality and dispersion of major crop distributions in the ten Southeast Asian countries, mean centers and standard deviation ellipses were calculated to analyze the spatial distribution of wheat, rice, and maize in 2020 (Fig. 2.12). A larger eccentricity of the standard deviation ellipse indicates higher spatial directionality and lower dispersion, and vice versa. The spatial distribution of wheat exhibited a southwest-to-northeast orientation, with the smallest spatial areas among the three crops. The mean centers of wheat was located in Southern Myanmar, and the ellipse of wheat displayed a relatively high eccentricity, indicating strong spatial directionality. Rice showed a broader spatial distribution, with its mean center located in Southern Vietnam and a northwest-to-southeast orientation. Maize exhibited the widest spatial distribution, and the mean center was in central Southeast Asia. The spatial dispersion of maize was the largest, and the directionality was the weakest (Fig. 2.12).

2. Per Capita Harvested Area and Total Production of Crops in the Ten Southeast Asian Countries in 2020

Figure 2.13 illustrates the per capita harvested area and total production of crops in the ten Southeast Asian countries in 2020. In most regions of Southeast Asia, the per capita harvested area of crops was less than 0.3 hm^2. However, in the northwestern regions, such as Myanmar, Thailand, Cambodia, and Vietnam,

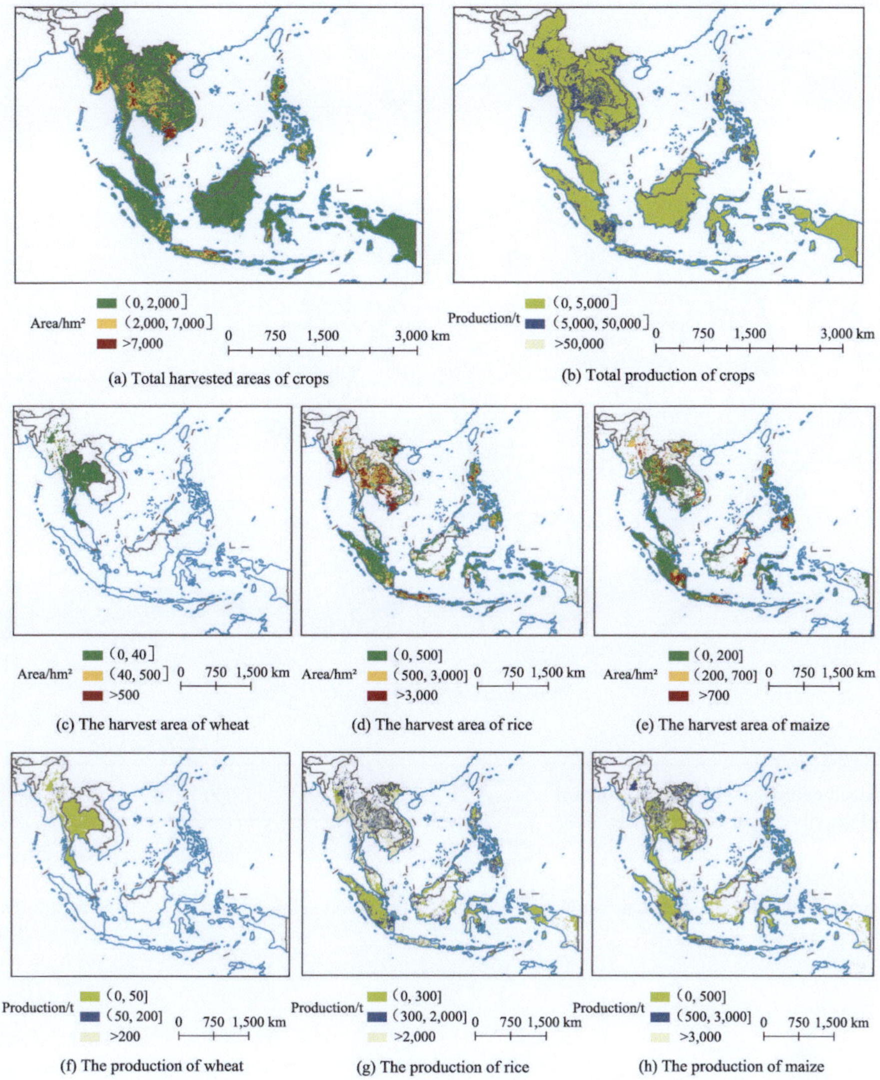

Fig. 2.11 Distribution of harvested areas and total production of all crops and major crops in the ten Southeast Asian countries in 2020

a significant number of pixels indicated the per capita harvested area of crops ranging from 0.3 to 1.5 hm^2 (Fig. 2.13a). These four countries, with abundant rainfall, a humid and hot climate, and fertile and flat land, provide suitable natural conditions for rice cultivation. As a result, they are the only countries in Southeast Asia that can maintain a net export of rice. Furthermore, government policies such as land reforms, the promotion of high-production varieties, and the adoption of advanced agricultural techniques have ensured stable rice production. The spatial distribution of per capita crop production was similar to that of the crop harvested area. In the majority of regions outside the northwestern part of Southeast Asia, the per capita crop production was generally less than 400 kg. The challenges associated with food production differ significantly across the ten Southeast Asian countries. In Singapore, limited land availability and environmental constraints pose substantial barriers to food production. Although Indonesia's climate and land conditions are favorable for agriculture, rapid population growth, imbalanced agricultural

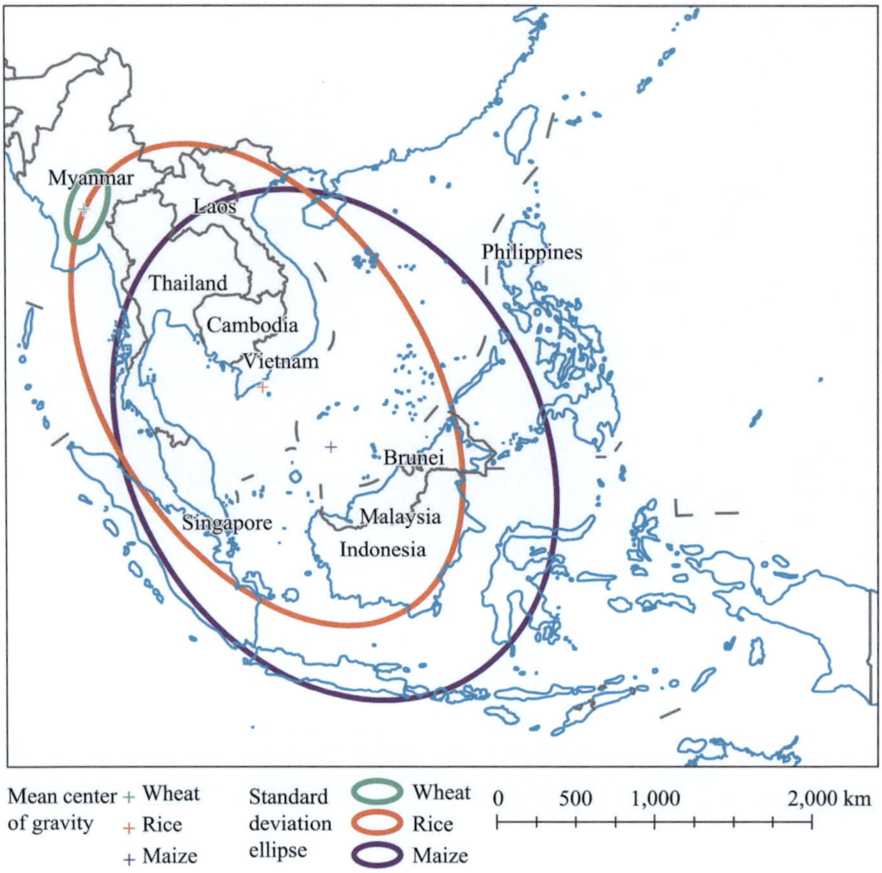

Fig. 2.12 Spatial distribution of mean centers and standard deviation ellipses for major crops in the ten Southeast Asian countries in 2020

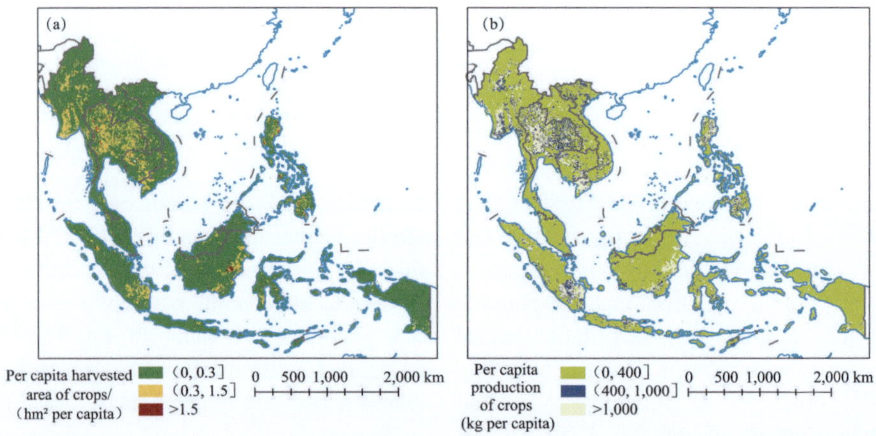

Fig. 2.13 Distribution map of per capita harvested area and per capita production of crops in the ten Southeast Asian countries in 2020

structures, outdated agricultural facilities, and frequent natural disasters severely threaten its food security. In Malaysia, the prioritization of economic growth and the expansion of cash crop cultivation have significantly reduced the land available for crop production.

3. **Trends in Total Production and Per Capita Production of Crops in the Ten Southeast Asian Countries from 2010 to 2020**

Based on the SPAM production data and population data from 2010 and 2020, the spatial distribution results of the trends in total crop production and per capita crop production in the ten Southeast Asian countries were obtained (Fig. 2.14). Among the major agricultural countries in Southeast Asia, Thailand and Vietnam showed a higher proportion of regions with an upward trend in total production due to favorable natural conditions and government policies. In contrast, Myanmar, Laos, and Cambodia showed a higher proportion of regions with a downward trend in total production. Notably, in recent years, the Philippines has been placing increasing emphasis on agricultural modernization, and most regions are maintaining stable or upward trends in total crop production. However, with the rising population in Southeast Asia from 2010 to 2020 and the decline in total crop production in some areas, the per capita crop production in most regions of Southeast Asia also showed a downward trend during this period (Fig. 2.14b), with rising trends primarily found in the Philippines, Thailand, and Central and Southern Indonesia.

Highlights

- A dataset on the spatial distribution of areas and yields of multiple crop types in Southeast Asian countries in 2020 was generated by fusing multiple sources of information such as remote sensing data and statistical data. This dataset contains the spatial distribution of areas and yields covering a wide range of crop types, which is useful for quantitative analysis of food production in Southeast Asia in 2020.
- A refined spatial grid representation of information on the state of food production in Southeast Asia was implemented by combining pixel-scale SPAM data and population data. This allows for a more detailed characterization of the spatial differentiation of food production patterns and provides an

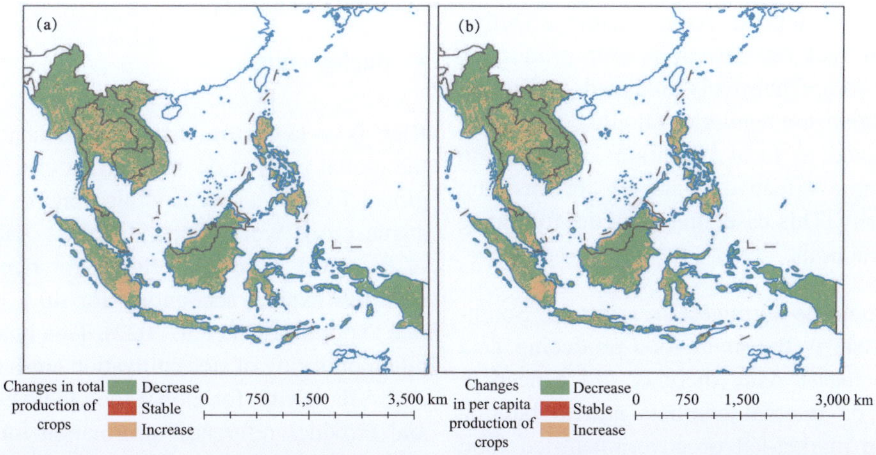

Fig. 2.14 Changes in total crop production and per capita crop production in the ten Southeast Asian countries from 2010 to 2020. *Note* The change rate $\leq -1\%$ is defined as a decrease trend, the change rate between -1% and 1% is defined as a stable trend, and the change rate $\geq 1\%$ is defined as an increase trend

important scientific basis for locally tailored food security policy formulation.

- **Discussion and Outlook**

This case study integrates multi-source remote sensing and statistical data, and utilizes SPAM to produce datasets of harvested areas and yields of multiple crops in the ten Southeast Asian countries in 2010 and 2020, which can provide important scientific data for national-scale SDG 2.4 assessment. First, in 2020, the total harvested area of crops in the ten Southeast Asian countries was mainly less than 2,000 hm^2, and the total production was mainly less than 5,000 t. Rice was the food crop with the largest harvested area in the Southeast Asian countries. Second, in most parts of Southeast Asia in 2020, the per capita harvested area for crops was less than 0.3 hm^2 and the per capita production of crops was less than 400 kg. Additionally, among the ten Southeast Asian countries from 2010 to 2020, Thailand, Vietnam, and the Philippines showed a clear increasing trend in total and per capita crop production, while the rest of the region showed a decreasing trend in food production.

Because of the significant spatial heterogeneity in harvested areas and crop yields resulting from different socio-economic and geographic conditions and planting patterns, the capacity to implement policies varies considerably across countries in Southeast Asia. Factors affecting changes in total and per capita crop production should be comprehensively analyzed, and scientific protection and regulation should be strengthened, tailored to local conditions, in order to develop more effective strategies for ensuring food security. This case study has the following specific proposals.

(1) Although the northwestern region of Southeast Asia is the main food-producing area in Southeast Asia, there is still a need to focus on internal conflicts and to build a mature market-led or government-led food reserve system to guarantee the stability of food production.

(2) The central region of Southeast Asia has a better base of natural conditions for agricultural production, and scientific reclamation or utilization of idle land should be encouraged. For example, consideration could be given to planting maize that can be harvested more quickly in palm plantations, investing more in agricultural infrastructure, increasing training in agricultural techniques, and breeding good varieties.

(3) For countries with limited natural conditions and territorial environments, it is important to develop modern urban agriculture, enact food-saving legislation, ensure the diversification of importing countries, and maintain a secure and stable food trade.

2.3.3 Extraction and Analysis of Interannual Rice Cultivation Areas in Southeast Asia

Target: SDG 2.4 By 2030, ensure sustainable food production systems and implement resilient agricultural practices that increase productivity and production, that help maintain ecosystems, that strengthen capacity for adaptation to climate change, extreme weather, drought, flooding and other disasters and that progressively improve land and soil quality.

- **Background**

Rice is a key crop that feeds nearly 50% of the global population and plays a crucial role in global food security. It is also one of the main staple crops worldwide. Southeast Asia is one of the most important regions for rice production and export, accounting for 40% of global rice exports (Yuan et al. 2022). Timely and accurate monitoring of rice cultivation areas in Southeast Asia is vital for global food yield estimation and provides reference information for government policy-making. This is especially significant for achieving SDG 2.4, which aims to establish sustainable food production systems.

Rice distribution information is critical, and multi-source remote sensing data have shown great potential for rice mapping. However, obtaining and updating global-scale, high-resolution rice spatial distribution information remains a challenging task. Currently, rice cultivation area mapping at the national or subcontinental levels typically relies on medium-resolution optical remote sensing data from instruments such as moderate resolution imaging spectroradiometer (MODIS) or Landsat. However, in Southeast Asia, cloud cover and rainfall often interfere with the acquisition of stable, continuous optical imagery, and the lower spatial resolution of such data is inadequate for the precise monitoring of fragmented and scattered rice fields (Dong et al. 2015; Shew and Ghosh 2019).

Methods for mapping rice cultivation areas based on time-series remote sensing data are generally divided into two categories: the phenology-based approach and the time-series data combined with the classifier-based approach. The phenology-based approach involves defining phenological indicators or combining time-series data with growth phenology calendars to determine specific phenological stages for rice extraction. While this approach is simple and computationally efficient, it relies heavily on the choice of phenological features and thresholds, which introduces uncertainty. The time-series data combined with the classifier-based approach directly uses time-series data as input features for machine learning or deep learning models, capturing fixed temporal relationships in the rice growth curve. Compared to temperate regions with stable single or double cropping systems, Southeast Asia has a complex rice growing cycle due to its humid, hot, and rainy climate. This flexibility in transplanting and sowing times makes it difficult to establish representative rice growth models for the region (Xu et al. 2021; Kang et al. 2022). This complexity presents challenges for both the phenological indicator-based and time-series model-based methods for rice mapping in Southeast Asia.

In contrast to optical remote sensing, synthetic aperture radar (SAR) is unaffected by cloud, fog, or rain, and is highly sensitive to the geometric structures and dielectric properties of crops. This makes SAR an ideal tool for all-weather and all-time monitoring of crops, especially for regions like Southeast Asia. To address the challenges posed by the complex rice cultivation patterns in tropical and subtropical regions, this case study utilizes Sentinel-1 time-series SAR data. A new mapping methodology was proposed that is suitable for large-scale tropical and subtropical regions. Specifically, by exploiting multi-scale spatiotemporal features of the Sentinel-1 time-series SAR data and incorporating prior knowledge about rice growth phenology and spatial distribution in Southeast Asia, an effective combination of SAR temporal statistical features was identified to capture key information about the rice growth process. The U-Net model was used to extract interannual rice cultivation areas for the period 2019–2021 in five Southeast Asian countries (e.g., Thailand, Myanmar, Vietnam, Cambodia, and Laos), generating a high-resolution, 20 m spatial resolution dataset of rice cultivation areas. This method provides a new solution for large-scale, precise rice mapping in Southeast Asia.

- **Data**
  - 2019–2021 Sentinel-1 SAR ground range detected (GRD) data, provided by the European Space Agency (ESA), with a spatial resolution of 22 m × 20 m (azimuth × range).
  - Shuttle radar topography mission (SRTM) 3 s digital elevation model (DEM), provided by NASA, with a spatial resolution of 90 m.
  - 2020 global land cover data WorldCover, provided by ESA, with a spatial resolution of 10 m.

- **Method**

This case study utilized the temporal statistical features of time-series Sentinel-1 vertical-horizontal (VH) polarization data and the U-Net model to perform rice cultivation area mapping for large-scale tropical or subtropical regions.

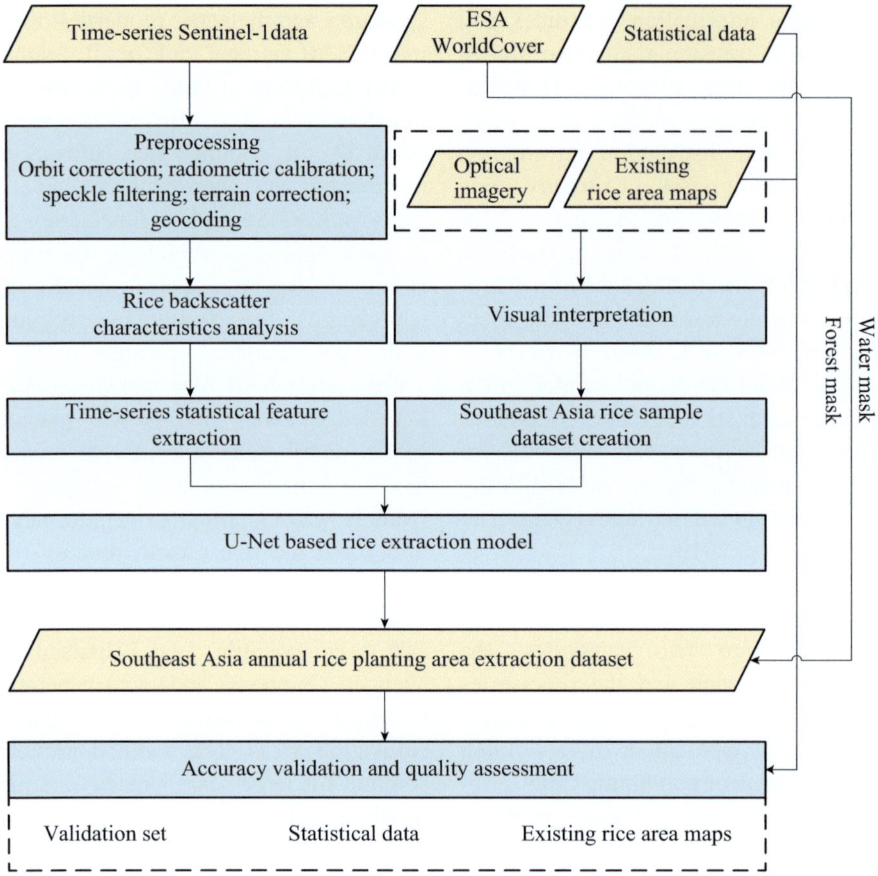

Fig. 2.15 Technical flowchart

A technical flowchart is shown in Fig. 2.15 and mainly includes the following steps: First, 7,880 scenes of time-series Sentinel-1 images from 2019 to 2021 were acquired, and VH polarization backscatter coefficients were obtained through data preprocessing. The data covers 90 scenes spatially, fully covering the five Southeast Asian countries, and temporally, each scene was observed 23–32 times, covering the entire time range during which rice might grow. Next, based on optical remote sensing images and other published rice cultivation area products, combined with visual interpretation, a rice sample set for Southeast Asia was built. By analyzing the backscatter temporal variation patterns of rice samples in Southeast Asia, key temporal statistical feature combinations for rice growth processes were extracted from multi-temporal SAR data. These features include temporal maximum, temporal minimum, and temporal variance. Among them, the temporal variance was used to distinguish cultivated land from other land cover types, the temporal minimum was used to identify the flooding period to differentiate rice from other crops, and the temporal maximum was used to reduce false alarms caused by water bodies. Based on the U-Net model, deep spatial information of rice was mined from the temporal statistical feature combinations to obtain rice cultivation area extraction results with fine spatial details. Finally, the results were refined using a mask based on the high-precision land cover product ESA WorldCover to reduce false alarms from water bodies and forests, resulting in an interannual rice cultivation area extraction dataset at a 20 m spatial resolution for the five countries in Southeast Asia from 2019 to 2021.

This case study used various strategies for accuracy verification. On the one hand, rice extraction accuracy was calculated based on the validation set; on the other hand, the dataset produced in this case study was evaluated for consistency with official statistical data and other rice products at both the national and provincial levels. The results show that this dataset has high accuracy and reliability. The verification data used are as follows.

(1) 2019–2020 Vietnam annual land use/land cover dataset, provided by the Japan Aerospace Exploration Agency (JAXA), with a spatial resolution of 30 m.
(2) 2000–2012 Asian rice data, provided by the International Rice Research Institute (IRRI), with a spatial resolution of 500 m.
(3) 2019–2021 statistical yearbooks for Vietnam, Cambodia, Laos, Thailand, and Myanmar, provided by the national statistical bureaus of each country.

- **Results and Analysis**

1. **Dataset Accuracy Verification**

Figure 2.16 shows the rice cultivation area map for the five Southeast Asian countries from 2019 to 2021, as obtained in this case study. The countries are ranked by total rice cultivation areas in descending order: Thailand, Myanmar, Vietnam, Cambodia, and Laos. In Thailand, rice cultivation is primarily concentrated in the central plains, northern, and northeastern regions. Myanmar's main rice-producing areas are located in the Irrawaddy delta, Bago region, and Yangon deltas, which are traversed by river systems. Additionally, dry northern regions, including Mandalay, Shwebo, and Magway, also play significant roles in rice production. In Vietnam, the most representative rice cultivation areas are the Mekong River Delta and the Red River Delta. Cambodia's rice production is mainly concentrated in the Tonle Sap Lake Basin and the southern Mekong River Basin. In Laos, the primary rice cultivation areas are located in the lowland regions of central and southern Laos. Many of the major rice-producing provinces are situated along the Mekong River, including Borikhamxay, Khammouane, Savannakhet, Salavan, and Champasak.

This case study comprehensively utilized various auxiliary data to validate the rice cultivation area extraction results for Southeast Asia. First, based on Google Earth optical imagery and other previously published rice cultivation area products, a rice accuracy validation sample set was constructed. Using this validation sample set, accuracy metrics for the rice cultivation area in the five Southeast Asian countries in 2019 were calculated. The accuracy was 92.20%, precision was 92.45%, recall was 90.26%, and the Kappa value was 0.8425. These results demonstrate that the large-scale rice extraction method proposed in this case study exhibits strong performance in rice extraction.

Furthermore, official statistical data from the five Southeast Asian countries and rice statistics from the IRRI were obtained. At both the national and provincial levels, the 20 m spatial resolution rice cultivation area extraction dataset produced in this case study was compared and analyzed with the official statistical data. Figure 2.17 shows the correlation between this case study's product and the statistical data at the national level. It can be observed that the R^2 values of this case study's product and the statistical data range from 0.87 to 0.92, indicating a high level of consistency between the two.

At the provincial (or administrative) scale, the correlation between the rice extraction results from this case study and the official statistical data as well as the IRRI rice data was calculated. Figure 2.18 shows, using Thailand as an example, the comparison of rice cultivation areas in 77 provinces with the official statistical data and

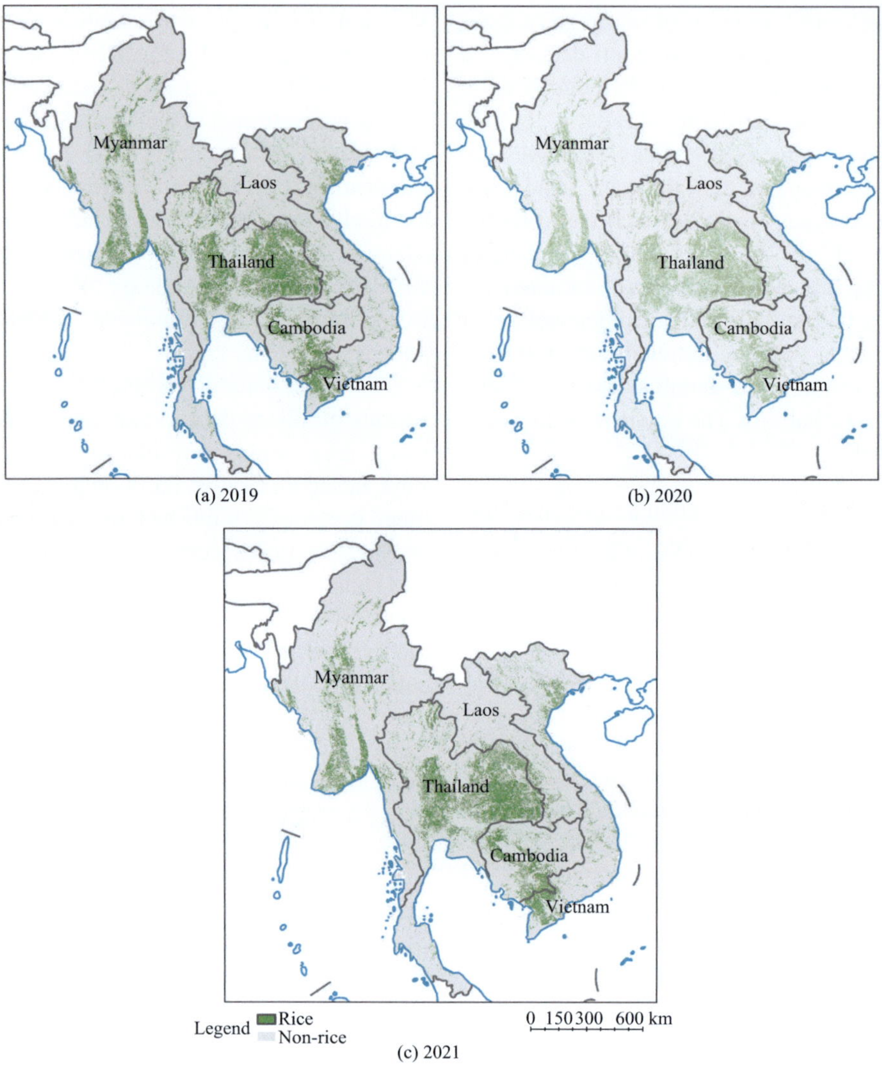

Fig. 2.16 Rice cultivation areas in the five Southeast Asian countries (2019–2021)

IRRI data. The R^2 values between this case study and the official statistical data range from 0.88 to 0.95, while those between this case study and the IRRI data range from 0.83 to 0.90. Additionally, for Vietnam, Myanmar, Cambodia, and Laos, the R^2 values at the provincial scale between this case study and both the statistical data and IRRI data are all above 0.8, indicating a high degree of correlation.

Furthermore, Fig. 2.19 compares the rice cultivation area proportions obtained from this case study with the statistical data from the FAO (FAOSTAT 2023), showing a high level of consistency. Based on the above analyses, it can be concluded that the interannual rice cultivation dataset at a 20 m spatial resolution for Southeast Asia from 2019 to 2021 had high reliability.

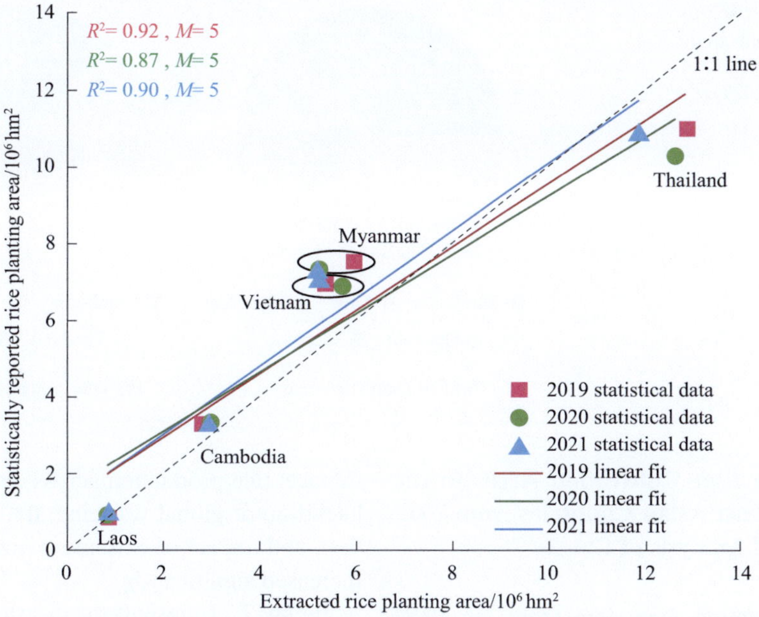

Fig. 2.17 Consistency analysis of the interannual rice cultivation areas and statistical data at the national scale for the five Southeast Asian countries (2019–2021)

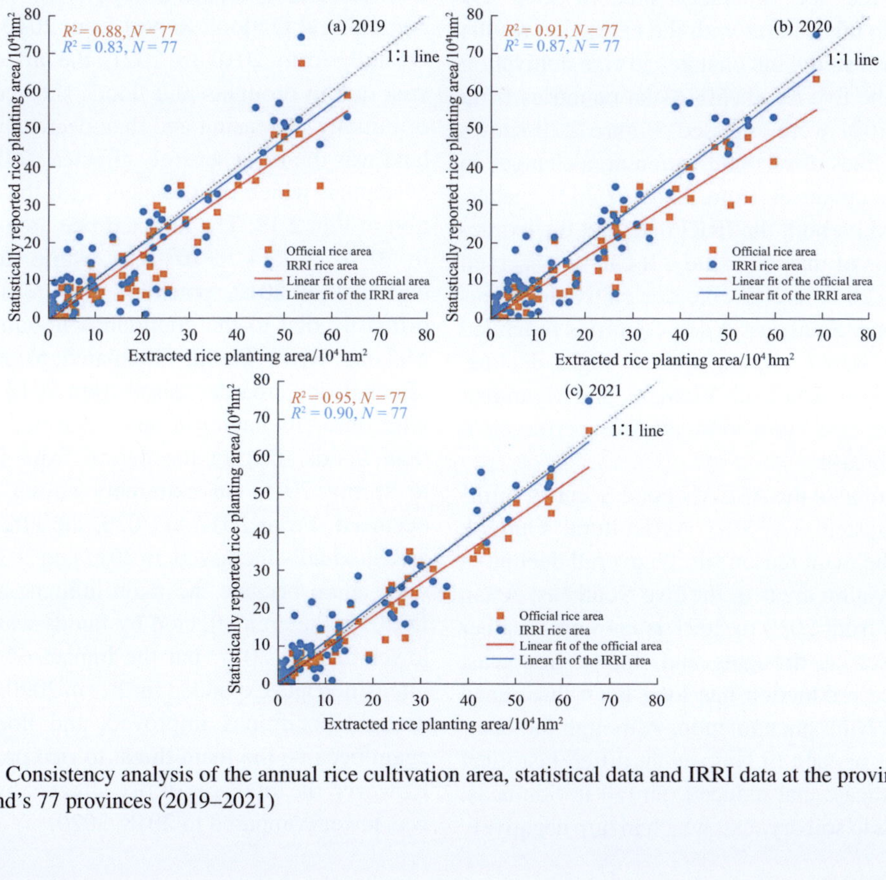

Fig. 2.18 Consistency analysis of the annual rice cultivation area, statistical data and IRRI data at the provincial scale for Thailand's 77 provinces (2019–2021)

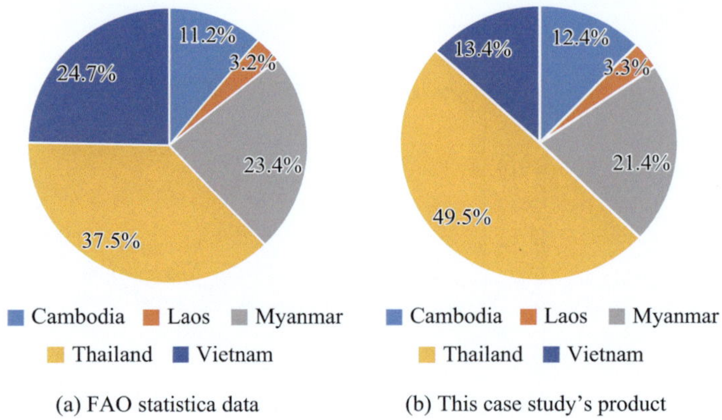

Fig. 2.19 Comparison of rice cultivation areas in the five Southeast Asian countries (2019–2021)

2. **Changes in Rice Cultivation Areas in the Five Southeast Asian Countries from 2012 to 2021 and Analysis of Causes**

The rice cultivation area data from the IRRI reflect the overall distribution of rice in Asia from 2000 to 2012, which can serve as a reference for the rice cultivation area in 2012. By comparing this dataset with the extraction results of this case study, the changes in rice cultivation areas in the five Southeast Asian countries from 2012 to 2021 were assessed. Figure 2.20 shows the trend lines of rice cultivation area changes in these five countries from 2012 to 2021, generated based on both the IRRI data and the extraction results of this case study. It can be observed that, from 2012 to 2021, the rice cultivation areas in these five Southeast Asian countries remained relatively stable overall, with a slight decline, particularly in Thailand, Vietnam, and Myanmar, where the rice cultivation areas experienced a minor decrease.

According to the ASEAN Food Security Information System (AFSIS) Agricultural Outlook Report, the main reason for the overall decline in rice cultivation areas in the five Southeast Asian countries from 2010 to 2021 is extreme weather events such as droughts and floods. Southeast Asian rice production has long been threatened by the El Niño phenomenon. Although the characteristics of each El Niño event differ, historical trends indicate that reduced rainfall in Southeast Asia leads to soil dryness, which in turn negatively impacts rice production across Asia. Against the backdrop of global warming, the intensity, duration, and impact of extreme weather events have increased significantly.

Figure 2.21 presents the disaster area statistics for rice in four Southeast Asian countries (Thailand, Myanmar, Cambodia, and Vietnam) affected by floods and droughts, as reported in the AFSIS Agricultural Outlook Report from 2010 to 2021. Overall, from 2010 to 2021, the affected rice area due to droughts and floods showed a trend of initially increasing and then decreasing. Thailand was the most severely affected, followed by Myanmar, which is consistent with the information in Fig. 2.19. The affected rice area occurred in 2010, at 2.14×10^5 hm^2, while the largest occurred in 2012, reaching 16.6×10^5 hm^2, primarily due to the continuous flooding in the Mekong River Basin. Compared to 2012, the affected rice area decreased from 2013 to 2016, with droughts having a more significant impact than floods. During the period from fall 2014 to spring 2016, an extremely strong El Niño occurred. From 2017 to 2020, the affected rice area gradually increased. In 2017 and 2018, floods once again became the main influencing factor. In 2019, the area affected by floods was slightly lower than in 2017, but the impact of droughts intensified and expanded further in 2020. In 2021, drought conditions improved, and floods once again became the main threat to rice production. However, the total area of rice affected by disaster was lower compared to 2018–2020.

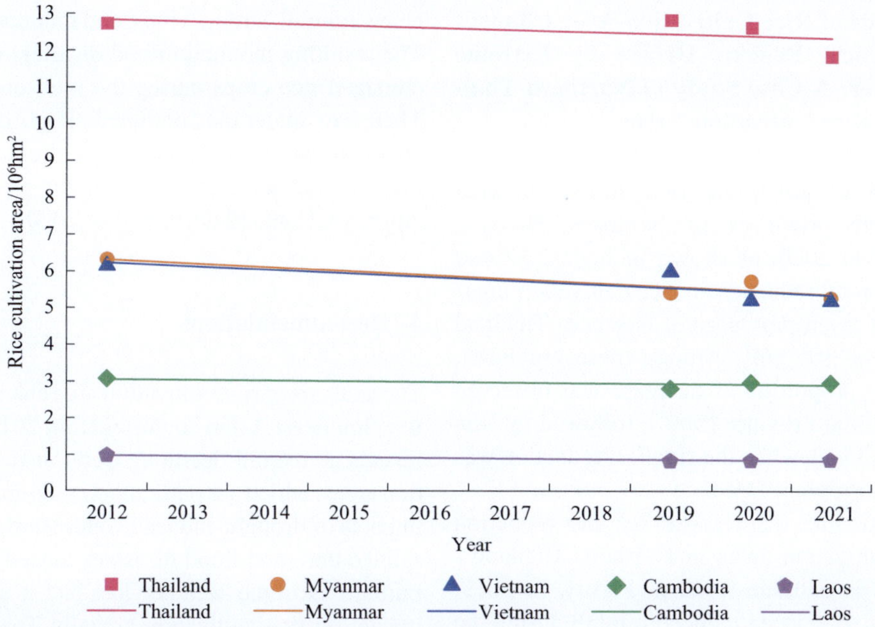

Fig. 2.20 Trends in rice cultivation areas in the five Southeast Asian countries (2012–2021)

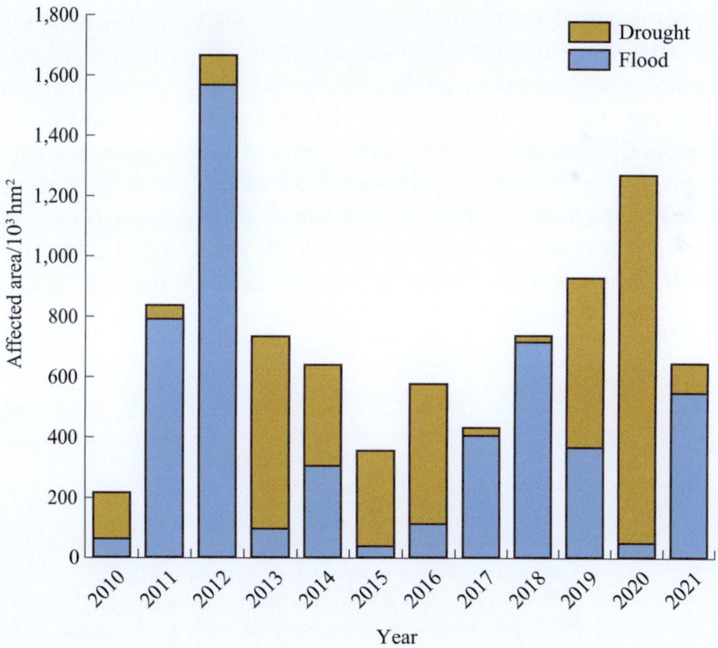

Fig. 2.21 Statistics of rice disaster areas affected by floods and droughts in four Southeast Asian countries (2010–2021)

3. **Analysis of Rice Cultivation Area Changes in Typical Regions Driven by Extreme Weather: A Case Study of Northern Thailand's Rice cultivation Areas**

Compared to 2019, the rice cultivation area in multiple provinces of Northern Thailand decreased in 2020, as shown in Figs. 2.22 and 2.23. The total reduction in rice cultivation areas across the seven provinces of Northern Thailand was 2.65×10^5 hm^2. Among these provinces, the largest proportion of decrease was observed in Chiang Mai Province (56%), followed by Nan Province (35%), while the smallest reduction was in Phayao Province (1%).

The possible main cause for the reduction in rice cultivation areas in Northern Thailand's provinces is extreme weather. Firstly, in 2019, Thailand experienced reduced rainfall during the monsoon season due to El Niño phenomenon, resulting in insufficient irrigation water. Secondly, the seasonal drought in 2020 (from January to April) occurred more rapidly and severely than in 2019. The shortage of irrigation water and natural water resources likely contributed to a reduction in the rice harvested area during the dry season. Compared to other regions, Northern and Northeastern Thailand experienced more severe drought conditions. Additionally, in August 2020, several provinces in Northern Thailand were hit by tropical storm "Sinlaku", which caused heavy rainfall, strong winds, and mountain floods. The resulting mountain flood disasters may have damaged rice crops during the monsoon season. Therefore, under the combined effects of drought and mountain floods induced by extreme weather, the rice cultivation areas in multiple provinces of Northern Thailand decreased in 2020.

4. **Recommendations**

The analysis of rice cultivation area changes in the five Southeast Asian countries from 2012 to 2021 reveals an overall declining trend in rice cultivation areas, which are particularly vulnerable to the impacts of drought-induced water shortages, land salinization, and flood disasters caused by heavy rainfall. Extreme weather has had a significant impact on rice cultivation areas in Thailand and Myanmar, while the effect on Cambodia, Laos, and Vietnam has been relatively minor. Based on the rice cultivation area situation in the five Southeast Asian countries, it is recommended that major rice-producing regions in Southeast Asia improve irrigation conditions, implement timely and effective post-disaster recovery measures, develop farmer-friendly policies, and establish resilient rice production systems to mitigate the adverse impacts of climate change, extreme weather, and other disasters, such as droughts and floods, on rice cultivation areas.

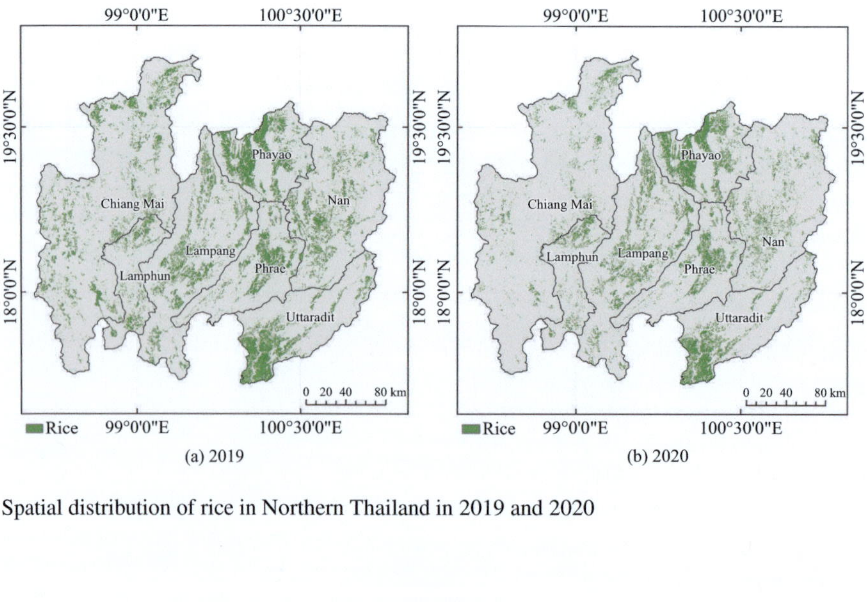

Fig. 2.22 Spatial distribution of rice in Northern Thailand in 2019 and 2020

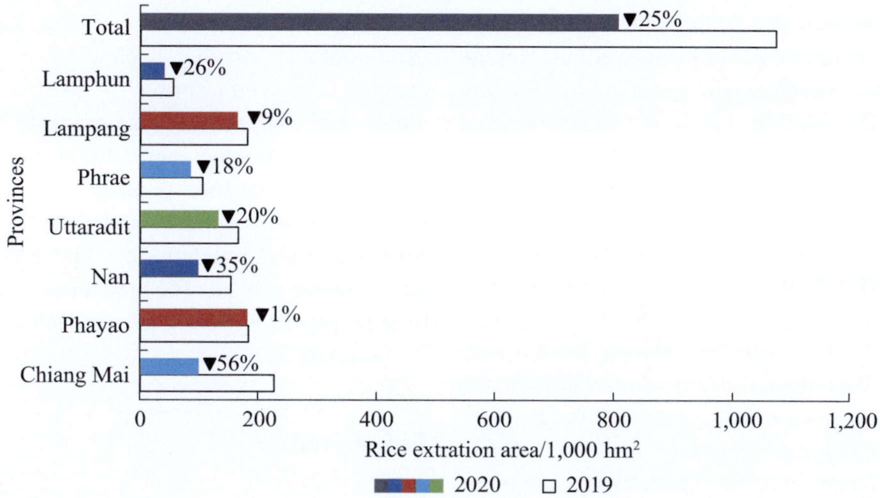

Fig. 2.23 Comparison of rice extraction areas in Northern Thailand's provinces (2019–2020)

Highlights

- A large-area tropical/subtropical SAR interannual rice mapping model based on temporal statistical features and the U-Net model is proposed. Three simple and effective combinations of statistical features, determined through temporal backscatter feature analysis, capture key information about the rice growth process. Combined with deep learning network models, this approach provides a new solution for rice mapping in tropical and subtropical regions.
- Using Sentinel-1 time-series SAR data and the proposed SAR rice mapping scheme, an interannual rice cultivation distribution product with a 20 m spatial resolution for the Southeast Asia region from 2019 to 2021 is generated, providing supporting data for sustainable agricultural assessments in Southeast Asia.

- **Discussion and Outlook**

This case study proposes a method for large-area tropical/subtropical rice cultivation mapping based on time-series SAR data. By analyzing the backscatter characteristics of Sentinel-1 time-series SAR data for rice in Southeast Asia, effective temporal statistical feature combinations are identified to capture key information about the rice growth process. The U-Net model is then used to extract interannual rice cultivation data for Southeast Asia from 2019 to 2021, with a 20 m spatial resolution. This dataset, along with additional information such as IRRI rice cultivation area maps and official statistical data, is used to analyze changes in rice cultivation areas in the five Southeast Asian countries from 2012 to 2021, providing data support for the establishment of a resilient global rice production system.

In the future, multi-source remote sensing data will continue to be used to monitor rice cultivation areas in Southeast Asia. On the one hand, the application of large-scale rice mapping methods will be further promoted across broader regions and longer time periods, enhancing the temporal and spatial scale of rice cultivation area products. This will support rice monitoring in tropical and subtropical regions and contribute to the development of a global sustainable food monitoring system. On the other hand, under shared socioeconomic pathways (SSP) and representative concentration pathways (RCP) scenarios, rice cultivation area change information will be integrated with rice yields, Southeast Asia's rice imports and exports, and international rice trade data. This will explore the role of climate change trends in rice production, identify key drivers of rice yield changes, and analyze the impacts of rice cultivation areas and yields on global food security. By focusing on extreme weather, droughts, floods, and other natural disasters, as well as climate change, this research will assist in formulating scientifically sound decisions on food reserves and trade, and provide better support for monitoring and evaluating SDG 2.4 indicators.

2.4 Summary

Globally, food security has always been a critical issue. The impacts of population growth and climate change have made ensuring food supply a focal point for governments and international organizations worldwide. This chapter integrates three case studies to demonstrate how Big Earth Data and remote sensing technologies can support the achievement of SDG 2.

Providing insights for agricultural policy formulation is crucial for promoting sustainable agricultural development and ensuring food security. Given the differences in agricultural policy implementation across countries and regions, region-specific recommendations are essential. In important rice production and export regions, this report generates datasets on the harvested areas and yields of multiple crops in the Southeast Asian countries by integrating multi-source remote sensing and statistical data. It reveals that from 2010 to 2020, the total harvested areas and yields of crops in the Southeast Asian countries showed an increasing trend, with spatial heterogeneity. Future efforts should focus on strengthening scientific protection and management to develop more effective strategies for enhancing global food production potential. Additionally, this report proposes a method for mapping rice cultivation areas based on time-series SAR data, which effectively captures key information during the rice growth process, thereby providing data support for establishing a resilient global rice production system. In the future, by combining rice yields, import and export volumes, and international trade information, multi-source remote sensing and statistical data can continue to monitor rice yields in Southeast Asia, thereby supporting decisions related to food reserves and trade.

The comprehensive use of Big Earth Data and remote sensing technology not only enables better monitoring of changes in crop yields but also provides a scientific basis for formulating effective agricultural policies and enhancing global food security. The application of these integrated studies and methods will help researchers and policymakers better understand and address global food security challenges, thereby providing references for achieving SDG 2 of the UN SDGs.

References

Becker-Reshef I, Barker B, Whitcraft A et al (2023) Crop type maps for operational global agricultural monitoring. Sci Data 10(1):172

Chen YQ, Li XB, Wang J (2011) Changes and effecting factors of grain production in China. Chin Geogra Sci 21(6):676–684

D'Amour CB, Reitsma F, Baiocchi G et al (2017) Future urban land expansion and implications for global croplands. Proc Natl Acad Sci USA 114(34):8939–8944

Dadrasi A, Chaichi M, Nehbandani A et al (2023) Global insight into understanding wheat yield and production through agro-ecological zoning. Sci Rep 13(1):15898

de S Nóia R Jr, Ewert F, Webber H et al (2022) Needed global wheat stock and crop management in response to the war in Ukraine. Glob Food Secur 35:100662

Dong JW, Xiao XM, Kou WL et al (2015) Tracking the dynamics of paddy rice planting area in 1986–2010 through time series Landsat images and phenology-based algorithms. Remote Sens Environ 160:99–113

Evans LT, Fischer RA (1999) Yield potential: its definition, measurement, and significance. Crop Sci 39(6):1544–1551

FAO (2021) The state of the world's land and water resources for food and agriculture: systems at breaking point. https://www.fao.org/land-water/solaw2021/en/

FAO, IFAD, UNICEF, et al (2023) The state of food security and nutrition in the world 2023: urbanization, agrifood systems transformation and healthy diets across the rural-urban continuum. https://openknowledge.fao.org/server/api/core/bitstreams/1f66b67b-1e45-45d1-b003-86162fd35dab/content

FAOSTAT (2023) Crops and livestock products[EB/OL]. https://www.fao.org/faostat/en/#data/QCL

Fischer RA (2015) Definitions and determination of crop yield, yield gaps, and of rates of change. Field Crop Res 182:9–18

Gerber JS, Ray DK, Makowski D et al (2024) Global spatially explicit yield gap time trends reveal regions at risk of future crop yield stagnation. Nat Food 5(2):125–135

Grogan D, Frolking S, Wisser D et al (2022) Global gridded crop harvested area, production, yield, and monthly physical area data *circa* 2015. Sci Data 9(1):15

Kang JM, Yang XM, Wang ZH et al (2022) Collaborative extraction of paddy planting areas with multi-source information based on google earth engine: a case study of Cambodia. Remote Sens 14(8):1823

Leng GY, Hall J (2019) Crop yield sensitivity of global major agricultural countries to droughts and the projected changes in the future. Sci Total Environ 654:811–821

Li YX, Zhang WF, Ma L et al (2014) An analysis of China's grain production: looking back and looking forward. Food Energy Secur 3(1):19–32

Licker R, Johnston M, Foley JA et al (2010) Mind the gap: how do climate and agricultural management explain the 'yield gap' of croplands around the world? Glob Ecol Biogeogr 19(6):769–782

Lipper L, DeFries R, Bizikova L (2020) Shedding light on the evidence blind spots confounding the multiple objectives of SDG 2. Nat Plants 6(10):1203–1210

Liu ZJ, Yang XG, Hubbard KG et al (2012) Maize potential yields and yield gaps in the changing climate of NorthEast China. Glob Change Biol 18(11):3441–3454

Mueller ND, Gerber JS, Johnston M et al (2012) Closing yield gaps through nutrient and water management. Nature 490(7419):254–257

Neumann K, Verburg PH, Stehfest E et al (2010) The yield gap of global grain production: a spatial analysis. Agric Syst 103(5):316–326

Portmann FT, Siebert S, Döll P (2010) MIRCA2000: global monthly irrigated and rainfed crop areas around the year 2000: a new high-resolution data set for agricultural and hydrological modeling. Glob Biogeochem Cycles 24(1):GB1011

Ray DK, Gerber JS, MacDonald GK et al (2015) Climate variation explains a third of global crop yield variability. Nat Commun 6:5989

Shew AM, Ghosh A (2019) Identifying dry-season rice-planting patterns in Bangladesh using the landsat archive. Remote Sensing 11(10):1235

United Nations (2022) The sustainable development goals report 2022. https://desapublications.un.org/publications/sustainable-development-goals-report-2022

Xiong W, Reynolds MP, Crossa J et al (2021) Increased ranking change in wheat breeding under climate change. Nat Plants 7(9):1207–1212

Xu L, Zhang H, Wang C et al (2021) Paddy rice mapping in Thailand using time-series sentinel-1 data and deep learning model. Remote Sens 13(19):3994

You LZ, Wood S, Wood-Sichra U et al (2014) Generating global crop distribution maps: from census to grid. Agric Syst 127:53–60

Yu QY, You LZ, Wood-Sichra U et al (2020) A cultivated planet in 2010–part 2: the global gridded agricultural-production maps. Earth Syst Sci Data 12(4):3545–3572

Yuan S, Stuart AM, Laborte AG et al (2022) Southeast Asia must narrow down the yield gap to continue to be a major rice bowl. Nat Food 3(3):217–226

Zabel F, Müller C, Elliott J et al (2021) Large potential for crop production adaptation depends on available future varieties. Glob Change Biol 27(16):3870–3882

Zhang TY, Yang XG, Wang HS et al (2014) Climatic and technological ceilings for Chinese rice stagnation based on yield gaps and yield trend pattern analysis. Glob Change Biol 20(4):1289–1298

Zhang TY, He Y, DePauw R et al (2022) Climate change may outpace current wheat breeding yield improvements in North America. Nat Commun 13(1):5591

3 SDG 6 Clean Water and Sanitation

3.1 Background

Based on the Summary Progress Update 2021: SDG 6—Water and Sanitation for All (UN-Water 2021), as of 2020, more than 3 billion people worldwide remain unaware of the quality of the rivers, lakes, and groundwater upon which they depend for survival. Between 2015 and 2020, global WUE improved by 9%, although significant spatial differences exist in this trend. Central and South Asia, East and Southeast Asia, and Oceania experienced the highest growth rates in WUE, while Latin America and the Caribbean saw declines. According to the latest UN assessment report (United Nations 2022), water-related ecosystems worldwide are degrading at an alarming rate. From 2015 to 2019, nearly one-fifth of the global basin surface water area underwent significant changes, including the expansion of water surfaces due to flooding and reservoir construction, as well as the loss of lakes, wetlands, and floodplains due to drought (UNEP 2021b).

However, analyses based on national statistical data are limited by the reporting cycles, resulting in low timeliness, and they fail to capture patterns at sub-national scales, which restricts their value for decision-making support. Additionally, the lack of standardized, comparable, and consistent monitoring data remains a long-term limiting factor in global and regional monitoring and assessment of three sub-targets: water quality, WUE, and changes in water-related ecosystems. In recent years, the use of satellite remote sensing image as a data source, combined with Big Earth Data analysis technology such as cloud computing and artificial intelligence, has enabled precise monitoring and measurement of global-scale surface water quality, WUE, and dynamic changes in water-related ecosystems. These remote sensing data products, developed based on technological advancements, not only fill the data gaps for corresponding indicators but also provide an effective approach for the large-scale, long-term monitoring of surface water resources, water quality, and WUE.

This chapter illustrates the feasibility and advantages of using Big Earth Data technology for constructing datasets on surface water transparency, agricultural WUE, and surface water distribution. Three global-scale case studies demonstrate their application in monitoring and evaluating surface water environments (SDG 6.3.2), WUE (SDG 6.4.1), and changes in water-related ecosystems (SDG 6.6.1).

3.2 Main Contributions

The main contributions of the three case studies in this chapter include the development of a global large lake and reservoir transparency retrieval model based on satellite remote sensing, the

H. Guo, *Big Earth Data in Support of the Sustainable Development Goals (2023)—The Belt and Road*, Sustainable Development Goals Series, https://doi.org/10.1007/978-981-95-3178-3_3

Table 3.1 Case and their main contributions

Targets	Tiers	Cases	Contributions
SDG 6.3 By 2030, improve water quality by reducing pollution, eliminating dumping and minimizing release of hazardous chemicals and materials, halving the proportion of untreated wastewater and substantially increasing recycling and safe reuse globally	Tier II	Global temporal and spatial changes in water transparency in large lakes	**Data product**: Produced a spatial distribution dataset of water transparency in global large lakes and reservoirs from 2000 to 2021 **Method and model**: Developed a global transparency retrieval model for large lakes and reservoirs based on satellite remote sensing
SDG 6.4 By 2030, substantially increase water-use efficiency across all sectors and ensure sustainable withdrawals and supply of freshwater to address water scarcity and substantially reduce the number of people suffering from water scarcity	Tier I	Changes in global cropland water-use efficiency	**Data product**: Global farmland dataset spanning 2001–2020
SDG 6.6 By 2020, protect and restore water-related ecosystems, including mountains, forests, wetlands, rivers, aquifers, and lakes	Tier I	Water area changes of global natural lakes and reservoirs	**Data product**: Triennial global surface water occurrence frequency dataset (2001–2021), tracking the distribution and dynamics of global lake and reservoir water bodies

spatial distribution of water transparency in large lakes and reservoirs worldwide from 2000 to 2021, and global water area changes of natural lakes and reservoirs alongside a triennial water occurrence dataset (Table 3.1).

3.3 Case Study

3.3.1 Global Temporal and Spatial Changes in Water Transparency in Large Lakes

Target: SDG 6.3 By 2030, improve water quality by reducing pollution, eliminating dumping and minimizing release of hazardous chemicals and materials, halving the proportion of untreated wastewater and substantially increasing recycling and safe reuse globally.

Indicator: SDG 6.3.2 Proportion of bodies of water with good ambient water quality.

- **Background**

Large lakes and reservoirs provide essential water resources, fishery resources, recreational venues, and significant ecological value to human society. They are a crucial component of the terrestrial hydrosphere and play a key role in global biogeochemical cycles. As sensitive indicators of global change, they have been termed the "sentinels" of global change (Ma et al. 2011; Williamson et al. 2009).

In recent decades, the intensification of threats such as rapid population growth, increased industrial and agricultural activities, and marked alterations in the hydrological cycle driven by climate change has led to a continuous increase in pollutant loads in lakes. Consequently, global lake water environments have undergone rapid changes, manifesting as more frequent harmful algal blooms, a decline in aquatic vegetation, a reduction in both the area and number of lakes, and a comprehensive decrease in water and food supply capacity (Woolway et al. 2020; Zhang et al. 2020).

A significant deficit exists in global-scale water quality data for lakes and reservoirs. The spatiotemporal variation of lake water quality, as well as its responses to climate change and human activities, remains an urgent research question in geography and remote sensing (Kavvada et al. 2020; Spyrakos et al. 2020; Tyler et al. 2016). The 2018 UN "Progress on Ambient Water Quality" report noted that, with respect to UN SDG 6, only 52 out of 193 UN member states have submitted surface water quality monitoring data, and in several cases, these data are based on a very limited number of monitoring stations (UN-Water 2018).

With the advancement of satellite remote sensing technology and the progress in theoretical methods for water color remote sensing, significant breakthroughs have been achieved in the optical remote sensing of inland waters in recent years (Duan et al. 2022; Zhang et al. 2021). Satellite remote sensing data are increasingly recognized as the most important and cost-effective source for large-scale surface water quality monitoring. In addition to filling critical data gaps, the ability of satellite data to provide extensive, long-term dynamic monitoring is a powerful tool for tracking global lake water quality trends.

Satellite remote sensing primarily monitors water quality parameters that influence the optical characteristics of water bodies, such as phytoplankton pigment concentrations, suspended sediment levels, colored dissolved organic matter concentrations, and water transparency. Among these, water transparency is a vital parameter for assessing water clarity and is central to understanding lake water environments, as its variations directly impact water security at both regional and global scales. Water transparency is determined by the combined effects of suspended sediments, phytoplankton pigments, and colored dissolved organic matter. Compared with other optical water quality parameters, water transparency more comprehensively reflects the overall water quality status, particularly in terms of turbidity. However, a universally applicable global model and long-term remote sensing inversion products for water transparency in lakes and reservoirs are still lacking, and research on the spatiotemporal changes in lake water transparency and its responses to environmental changes remains insufficient.

This case study leverages Big Earth Data technology and uses water transparency as an indicator for lake water quality monitoring. Based on satellite remote sensing analysis, it examines the spatiotemporal trends of water transparency in large lakes and reservoirs (with an area exceeding 25 km^2) worldwide from 2000 to 2021, thereby providing a novel global remote sensing dataset for lake water transparency to support the achievement of SDG 6.

- **Data**
  - Satellite remote sensing data: Terra MODIS data with a 500 m spatial resolution covering the globe from 2000 to 2021.
  - In-situ water data: Measured transparency dataset of China's surface waters (Wang et al. 2020); in-situ transparency datasets from the National Earth System Science Data Center and the China Lake Science Database; the shared dataset from the European Multi Lake Survey (EMLS); and the shared AquaSat dataset from the United States.
  - Basic geographic information: Global coastal vector data.

- **Method**

This case study used MODIS surface reflectance products as the primary data source to construct a surface water transparency inversion model based on Forel-Ule index (FUI) and hue angle (Wang et al. 2020). Field-measured water datasets collected from different regions worldwide were used to validate and calibrate the surface water transparency inversion model. Based on MODIS, a product set of water transparency for global large lakes and reservoirs with areas greater than 25 km^2 from 2000 to 2021 was generated. On this foundation, the spatial and temporal variation trends of water transparency in global large lakes and reservoirs from 2000 to 2021 were analyzed.

- **Results and Analysis**

1. **Global Spatial Pattern Analysis of Water Transparency in Large Lakes and Reservoirs from 2000 to 2021**

The distribution of global climate mean values of water transparency in large lakes and reservoirs from 2000 to 2021 is shown in Fig. 3.1, and the differences in the distribution of water transparency in different continents and climatic zones are shown in Fig. 3.2. From Fig. 3.1, it can be seen that the distribution of global water transparency in large lakes and reservoirs varies greatly by region, and the overall distribution is concave with latitude, i.e., the transparency of lakes and reservoirs near the North and South Poles (high latitude regions) is higher, while the transparency of lakes and reservoirs near the equator (low latitude regions) is lower. As can be seen from Fig. 3.2, in the global continents, from the point of view of the average water transparency, lake water transparency in Asia and Europe is higher, the lowest lake water transparency is in Africa. From the point of view of the number of large lakes and reservoirs, Asia and North America have the largest number of large lakes and reservoirs, while Oceania has the fewest. In different climatic zones of the world, from the point of view of average water

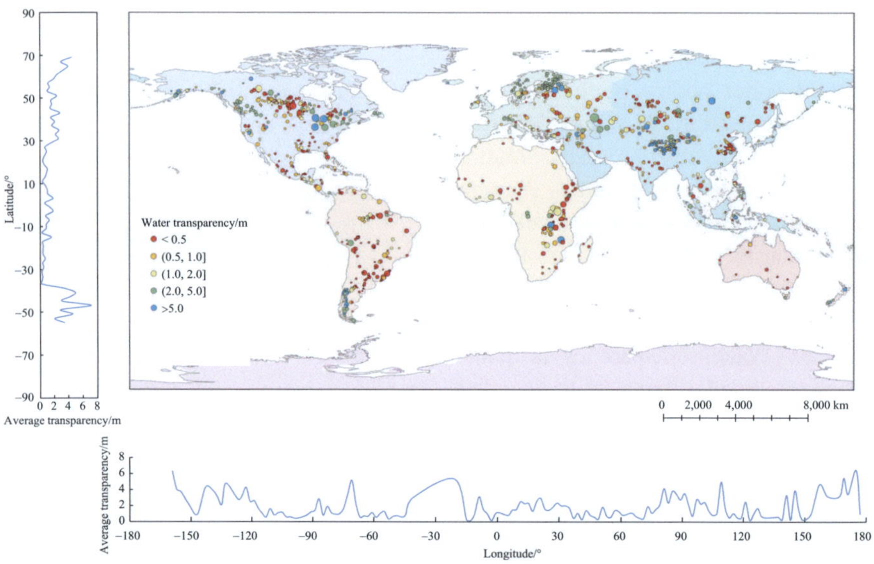

Fig. 3.1 Distribution of global climatic mean values of water transparency in large lakes and reservoirs from 2000 to 2021

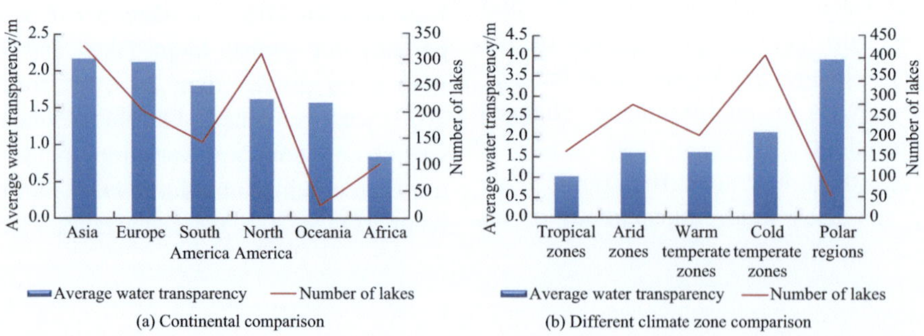

Fig. 3.2 Comparison of the average transparency and number of water bodies of large lakes in different continents and climatic zones from 2000 to 2021

transparency of lakes and reservoirs, the water transparency of lakes and reservoirs is higher in polar and cold temperate zones and lower in tropical zones. While from the point of view of quantity, the number of large lakes and reservoirs is higher in cold temperate zones and arid zones, and lowest in polar regions.

2. **Analysis of the Global Long-Term Time-Series Change in Water Transparency of Large Lakes and reservoirs from 2000 to 2021**

The annual change rate of water transparency of large lakes and Reservoirs in different climatic zones during 2000–2021 is depicted in Fig. 3.3, and the differences in the changes in different continents and climatic zones are shown in Figs. 3.4 and 3.5. The global change rate of water transparency in large lakes and Reservoirs showed an obvious regional distribution, with significant differences in regions and climatic zones, though an overall upward trend of water transparency has been obvious since 2000. The water transparency of 44.2% of large lakes and Reservoirs showed a significant upward trend ($p < 0.05$), and only 10.6% of large lakes and Reservoirs showed a significant downward trend in water transparency ($p < 0.05$). From the statistics of all continents, the average annual change rates of water transparency in large lakes and Reservoirs across the six continents were all positive, as shown in Fig. 3.4. Among them, the average water transparency of large lakes and Reservoirs in Asia and Africa changed slowly, with average annual change rates of 1.3 cm/a. The average water transparency of large lakes and Reservoirs in Europe increased significantly, with an average annual change rate of 7.6 cm/a. From the perspective of different climatic zones (Fig. 3.5), the water transparency of large lakes and Reservoirs in the cold temperate zones increased most obviously, with an average annual change rate of 5.6 cm/a, and the water transparency of polar large lakes and Reservoirs increased significantly, with an average annual change rate of 3.1 cm/a. The increase in water transparency in tropical and arid zones was weak, and the water transparency of turbid lakes ($Z_{SD} < 0.5$ m) in the warm temperate zone showed a downward trend.

Highlights

– This case study proposed and tested the satellite remote sensing-based inversion model for

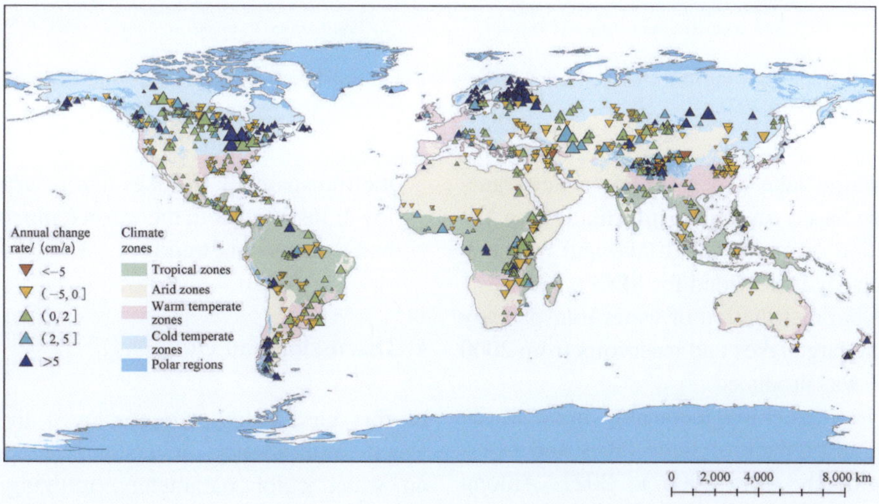

Fig. 3.3 Global annual change rates of water transparency in large lakes and Reservoirs in different climatic zones during 2000–2021

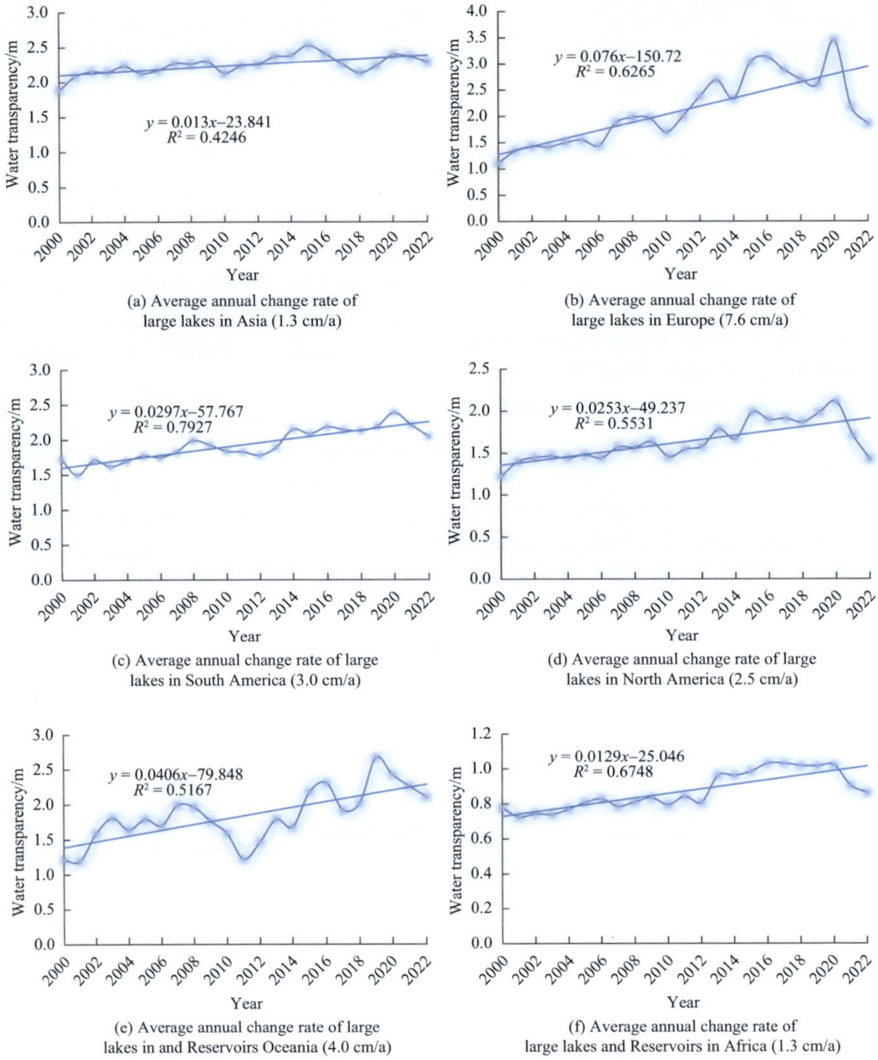

Fig. 3.4 Line graph showing the change in the average water transparency of large lakes and Reservoirs in each continent from 2000 to 2021

global large lake and reservoir water transparency, based on which information on the proportion of good global lake and reservoir water clarity is provided for SDG 6.3.2.

- The spatial distribution of water transparency of global large lakes and reservoirs from 2000 to 2021 was produced.
- There is a generally increasing trend in the water transparency of large lakes and reservoirs globally from 2000 to 2021. Among them, the transparency of lakes and reservoirs in the cold zones increased significantly, while the transparency of lakes and reservoirs with low transparency in the warm temperate zones had a decreasing trend.

- **Discussion and Outlook**

In this case study, we employed the surface water transparency inversion method based on water color parameters, utilizing satellite remote sensing data as the primary data source and ground-measured data from domestic and

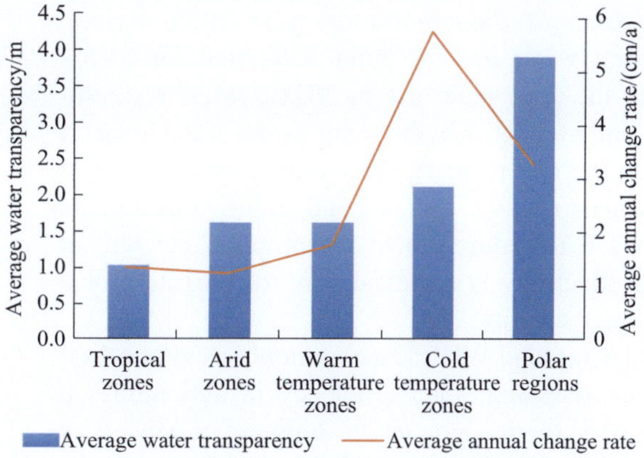

Fig. 3.5 Comparison of the average water transparency and annual change rate of large lakes and Reservoirs in different climatic zones

international lakes and reservoirs as auxiliary data. We constructed and validated the global water transparency model for large lakes and reservoirs, producing a dataset of water transparency for global large lakes and reservoirs with areas exceeding 25 km^2 during 2000–2021. Water transparency is used as an indicator of surface water quality to enable global monitoring of the SDG 6.6.1 indicator.

Analyses were conducted using continents and global climatic zones as geographical units, highlighting the differences in water transparency of large lakes and reservoirs across different continents and climatic zones, as well as their changes from 2000 to 2021. This case study provides crucial support for achieving SDG 6 through comprehensive data on spatiotemporal water quality changes and pattern analysis.

Currently, water quality parameters monitored through remote sensing for lakes and reservoirs are primarily optical parameters, such as the water transparency of large lakes and reservoirs monitored in this case study. In future work, we will further explore the relationships between optical water quality parameters and the biochemical indicator-based water quality parameters in SDG 6.3.2, to foster closer integration between remote sensing monitoring of lake and reservoir water quality and the SDG 6.3.2 indicator.

3.3.2 Changes in Global Cropland Water-Use Efficiency

Target: SDG 6.4 By 2030, substantially increase water-use efficiency across all sectors and ensure sustainable withdrawals and supply of freshwater to address water scarcity and substantially reduce the number of people suffering from water scarcity.

Indicator: SDG 6.4.1 change in water-use efficiency over time.

- **Background**

Improving WUE has always been a prominent topic across various industries, closely tied to human well-being and SDGs. In 2015, the UN introduced the SDGs, with SDG 6.4.1 specifically focusing on "change in water-use efficiency over time", aiming to measure the variations in water resource utilization efficiency across countries, encompassing sectors such as agriculture, industry, and service industry. Agriculture, characterized by substantial water usage and high water consumption, stands out as a major consumer of water among the three primary sectors—agriculture, industry, and service industry. According to the *China Water*

Resources Bulletin for the year 2021, agricultural water use in China accounted for 61.5% of the total water use in 2021, which is significantly higher than the water usage of the industry and service industry. With limited water resources and increasing water demand, improving cropland WUE to reduce water consumption per unit of productivity is a critical way to mitigate global water scarcity.

Cropland WUE is a commonly used indicator for assessing water efficiency in agriculture. It refers to the amount of biomass produced or the economic value generated per unit of water, reflecting the WUE and the economic benefits generated from the perspective of output. This indicator can comprehensively reflect the trade-off relationship between cultivated land food production and water consumption (Yu et al. 2008). Remote sensing technology has played a pivotal role in the timely, rapid, and large-scale estimation of regional cropland output, water consumption, and WUE. This technology is advantageous in elucidating the patterns of cropland WUE variations, contributing to global and regional sustainable utilization of agricultural water resources, and supporting sustainable agricultural development (Ai et al. 2020). However, there is currently a lack of high temporal and spatial resolution, long-time series datasets on cropland WUE (Cai et al. 2021). Therefore, this case study developed a global assessment methodology for cropland WUE based on multi-source data to generate a 1 km resolution dataset on cropland WUE from 2001 to 2020 and then evaluated the global patterns and temporal variations in cropland WUE, providing valuable data support for sustainable agricultural development and the sustainable utilization of water resources.

- **Data**
  - Fraction of photosynthetically active radiation (fPAR) and leaf area index data from the global land surface satellite (GLASS) product (8 days/250 m) (http://www.glass.umd.edu/Download.html). The dataset was resampled from a spatial resolution of 250 m to 1 km using the mean resampling method, and from a temporal resolution of 8 days to 1 day using linear interpolation.
  - Downward shortwave radiation, air temperature, and dew-point temperature data from the European Centre for Medium-Range Weather Forecasts (ECMWF) Fifth-Generation ECMWF Atmospheric Reanalysis of the Global Climate (ERA5) dataset (https://www.ecmwf.int/en/forecasts/datasets/reanalysis-datasets/era5). The dataset was resampled to 1 km spatial resolution using the statistical downscaling method.
  - Ground measurements of latent heat flux and CO_2 flux or gross primary productivity (GPP) from the global flux tower observed dataset (https://fluxnet.org).
  - Global land use/land cover datasets from the European Space Agency-Climate Change Initiative (ESA-CCI) (http://maps.elie.ucl.ac.be/CCI/viewer/download.php). The dataset was resampled from 300 m to 1 km spatial resolution using the majority resampling method.
  - High-resolution (1 km) Köppen-Geiger climate classification dataset (http://koeppen-geiger.vu-wien.ac.at/shifts.htm).

- **Method**

Cropland WUE was calculated as the ratio of net primary productivity (NPP) to evapotranspiration (ET) using the following equation:

$$\mathrm{WUE} = \frac{\mathrm{NPP}}{\mathrm{ET}}$$

where:

(1) Cropland NPP is crop GPP minus the fraction of the stored organic carbon consumed by crops to maintain crop respiration (CR). GPP was estimated using an improved evaporative fraction light-use-efficiency (EF-LUE) model (Du et al. 2022), which introduced a new parameterization scheme to account for the soil water stress factor, thus improving the accuracy of GPP estimation under water

stress conditions. The model parameters, such as maximum light-use efficiency, temperature stress, and vapor pressure deficit stress, were calibrated and validated based on GPP observations obtained from the global flux tower observed dataset for different climate regions. CR was obtained by summing the growth respiration (proportional to NPP at the annual scale) and maintenance respiration (related to the leaf area index, etc. at the daily scale, and accumulated to obtain the annual sum).

(2) ET was calculated by applying the ETMonitor model (Hu and Jia 2015; Zheng et al. 2019, 2022) to the corresponding multi-source remotely sensed data and atmospheric reanalysis dataset (ERA5). ETMonitor comprehensively considers the major physical processes affecting evapotranspiration, including energy balance, water balance, and vegetation physiology.

(3) Global cropland WUE product for 2001–2020 with an annual time step and a spatial resolution of 1 km was created using the aforementioned NPP and ET data. The validation results based on the global flux tower observed dataset indicate that the RMSE of the cropland WUE results produced by this case study is 0.5 g C/kg H_2O/a, which is considered to be of high precision.

- **Results and Analysis**

1. **Spatial Patterns and Changes in Global Cropland WUE**

The global average cropland WUE over multiple years (2001–2020) is approximately 0.98 g C/kg H_2O/a. There are noticeable spatial variations in the distribution of global cropland WUE. The pattern reveals that tropical regions exhibit the highest cropland WUE, with an average of approximately 1.13 g C/kg H_2O/a, followed by arid and semi-arid regions approximately 0.95 g C/kg H_2O/a, and cold regions approximately 0.95 g C/kg H_2O/a. Temperate regions, comparatively, have a relatively lower average of 0.88 g C/kg H_2O/a.

From 2001 to 2020, there was a significant increasing trend in cropland WUE in agricultural areas worldwide (Fig. 3.6), with an overall rise of 3.48%. However, spatial differences were observed. On the continental scale, differences were evident. Asia had the largest increase in cropland WUE (8.9%), followed by Oceania (7.2%). North America also showed an increasing trend, but with a lower rate (4.6%) compared to Asia and Oceania. Europe, Africa, and South America had relatively small changes in cropland WUE (all less than 1%). The global increase in cropland WUE is primarily attributed to the greater increase in agricultural NPP (9.31%) compared to the increase in evapotranspiration (6.39%). This is due to the significant advancements in agricultural technology and infrastructure, particularly in countries like China, as well as improvements in water-saving irrigation and agricultural water management techniques.

2. **Changes in Cropland WUE of Typical Countries**

To further clarify the changing characteristics of global cropland WUE, this case study selected typical countries from major grain-producing countries, and countries associated with the BRI and their neighboring countries in North Africa and Central Asia for cropland WUE analysis. Major grain-producing countries include China, the United States, Brazil, Indonesia, Argentina, France, and Canada. North African countries comprise Morocco, Algeria, Tunisia, Libya, and Egypt. Countries in Central Asia include Kazakhstan, Kyrgyzstan, Tajikistan, Uzbekistan, and Turkmenistan.

From 2001 to 2020, the cropland WUE of the world's major grain-producing countries showed an upward trend (Fig. 3.7). Canada and China experienced the highest increases, with growth rates of 18.2% and 13.3%, respectively. Indonesia, Brazil, and France, with higher average cropland WUE, had smaller increases, all below 2%. The United States and Argentina had increases of 3.2% and 2.6%, respectively. The significant improvement in China is attributed to the much larger increase

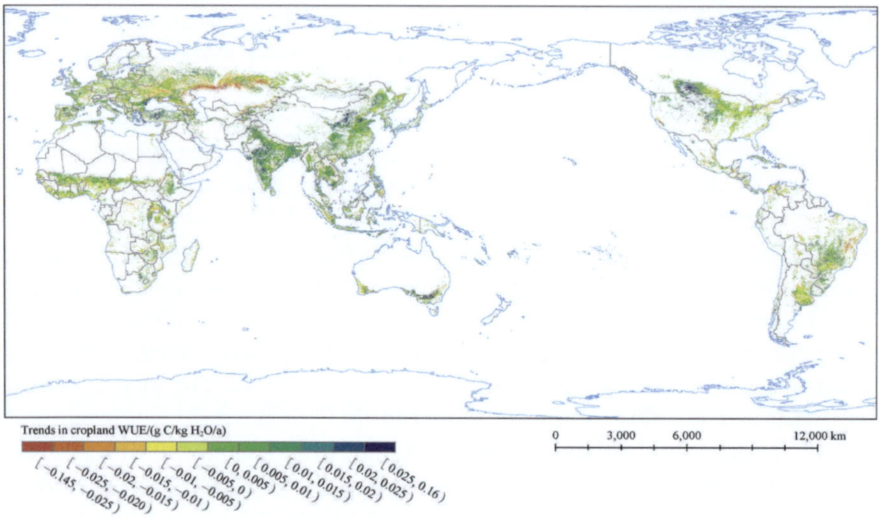

Fig. 3.6 Interannual trend map of global cropland WUE from 2001 to 2020

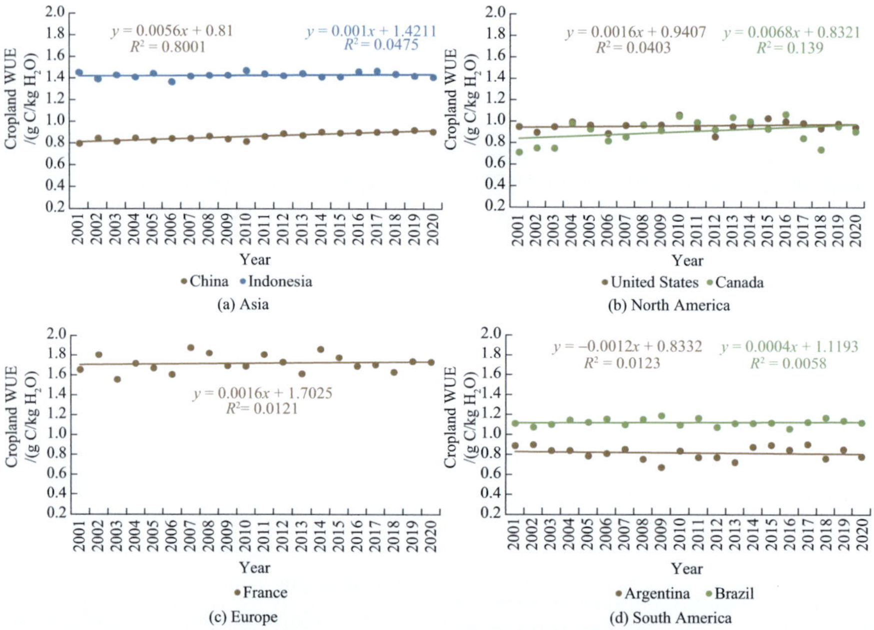

Fig. 3.7 Annual variation of cropland WUE from 2001 to 2020 in the major grain-producing countries worldwide

in cropland NPP (over 25%) compared to the increase in evapotranspiration (approximately 10%). This improvement is mainly due to factors such as advances in agricultural technology (e.g., field management, water-saving measures, fertilization, breeding, etc.), adjustments in cropping structure and intensity, and climate change (e.g., elevated CO_2 concentration, etc.) (Chen et al. 2019; Yang et al. 2022; Zhai et al. 2021). For example, from 2001 to 2021, due to advancements in agricultural technology, China's irrigation WUE continuously improved from 0.43 to 0.568, and the net water consumption per unit of grain production decreased from 0.469 m^3/kg to

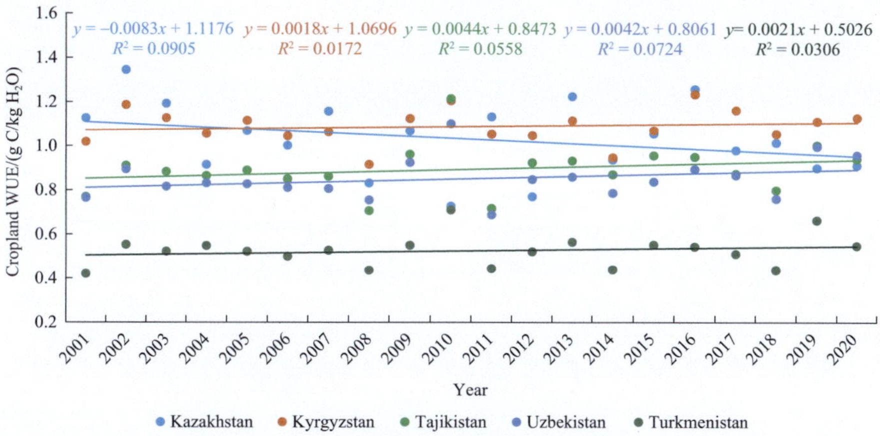

Fig. 3.8 Annual variation of cropland WUE from 2001 to 2020 in the five Central Asian countries

0.404 m^3/kg, with an average annual reduction of 0.003 m^3/kg (Kang 2022).

In Central Asia (Fig. 3.8), Turkmenistan, Tajikistan, and Uzbekistan, which have relatively low average cropland WUE, experienced an increasing trend with a significant growth of approximately 10.3%. Kyrgyzstan, with higher average cropland WUE, also showed an upward trend, but the increase was relatively small, at only 3.3%. Kazakhstan, however, experienced a decreasing trend with a substantial decrease of 13.9% (Fig. 3.8). The primary reason for the improvement in cropland WUE in Turkmenistan, Tajikistan, and Uzbekistan may be attributed to the adjustment of agricultural planting structures in irrigation areas. From 2001 to 2020, in response to food crises, a considerable number of cotton fields in irrigation areas were converted to winter wheat cultivation. Winter wheat has higher WUE compared to cotton, leading to an overall enhancement in WUE (Zou et al. 2017). In the case of Kazakhstan, its decreasing cropland WUE is likely due to drought conditions in the rainfed agricultural region in the northwest (Dubovyk et al. 2019; Karatayev et al. 2022), and land degradation exacerbated by human activities such as grassland cultivation and land abandonment (Hu et al. 2020). This collective impact significantly reduced the cropland NPP (by approximately 11.4%), contributing to the observed decline in cropland WUE.

In the North African countries (Fig. 3.9), Algeria, Libya, and Tunisia all showed an increasing trend in cropland WUE, with growth rates of 14.2%, 11.5%, and 7.5%, respectively. In contrast, Morocco and Egypt experienced a decreasing trend, with reductions of 8.9% and 9.9%, respectively. The rise in cropland WUE in Algeria, Libya, and Tunisia is primarily attributed to a significant increase in cropland NPP, far exceeding the increase in water consumption through ET. In the case of Morocco, the decrease in cropland WUE mainly results from a reduction in cropland NPP (approximately 4%) caused by decreased precipitation, and a slight increase in water consumption through ET (3%) due to rising temperatures. For Egypt, the substantial increase in ET (about 10%) is mainly due to a significant rise in temperatures, and a slight decrease in cropland NPP (a decrease of 0.7%), leading to a notable decrease in cropland WUE.

Highlights

– This case study developed a consistent and spatially comparable global dataset of cropland WUE from 2001 to 2020, providing methods and data for monitoring and assessing the SDG 6.4.1 indicator.
– This case study analyzed the spatiotemporal pattern of global cropland WUE, providing data support for the sustainable utilization of

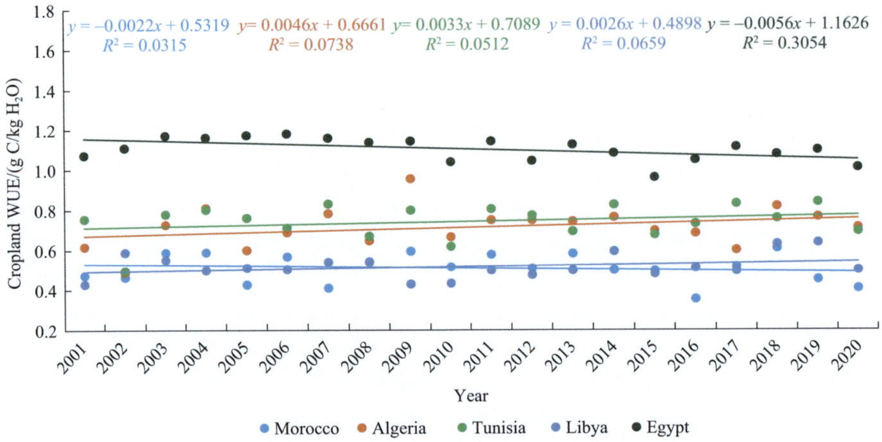

Fig. 3.9 Annual variation of cropland WUE from 2001 to 2020 in the five North African countries

agricultural water resources and the sustainable development of agriculture at both global and regional levels.

- **Discussion and Outlook**

This case study developed the global dataset of annual cropland WUE, defined as the ratio of cropland NPP to ET, at a 1 km resolution from 2001 to 2020, based on multi-source data and incorporating models. This case study assessed and analyzed the spatiotemporal variations in cropland WUE globally and in typical countries, providing data support for the evaluation of agricultural WUE and its time-series changes aligned with SDGs.

The results indicate that there was a significant increase in cropland WUE in agricultural areas worldwide, but spatial differences were observed. The global increase in cropland WUE is primarily attributed to significant advancements in agricultural technology, substantial improvements in agricultural infrastructure conditions, and the adoption of water-saving irrigation and agricultural water management technology, with countries like China leading the way. It is recommended that countries and regions with lower WUE should increase investments in agricultural technology and infrastructure to enhance cropland WUE, promoting sustainable water resource utilization. Furthermore, against the backdrop of climate change, different regions exhibit varied responses of cropland WUE to climate change. For regions where climate change leads to a decrease in cropland WUE, tailored measures are suggested to minimize the adverse effects of climate change (grain reduction, water consumption increase, etc.), considering the specific conditions of each area.

3.3.3 Water Area Changes of Global Natural Lakes and Reservoirs

Target: SDG 6.6 By 2020, protect and restore water-related ecosystems, including mountains, forests, wetlands, rivers, aquifers, and lakes.

Indicator: SDG 6.6.1 Change in the extent of water-related ecosystems over time.

- **Background**

Lakes, including natural lakes and manmade reservoirs, play vital roles in the global hydrological and biogeochemical cycles and underpin vital ecosystem functions and services. However, rapid lake changes have been identified worldwide in response to changing climate and escalating human activities, threatening the ecosystem services provided by these lacustrine habitats. A spatially explicit understanding of their extent changes is essential for evaluating the associated ecological, environmental, and societal impacts, which is also a key indicator within the UN SDG 6 framework (SDG 6.6.1). This particular indicator includes five sub-indicators, with this case study focusing on the changes in the spatial extent of both natural and artificial water bodies. Natural water bodies refer primarily to natural lakes, while artificial water bodies are designated as reservoirs developed with regulatory facilities like dams for water storage and management (Cooley et al. 2021; Yao et al. 2019). Specifically, human management practices have intensified seasonal variations in water storage, resulting in reservoirs exhibiting distinct ecohydrological characteristics compared to their natural counterparts (Cooley et al. 2021). Accurately distinguishing these two categories is essential for evaluating the available water resources for humans. The identification and extraction of natural lakes and reservoirs through Big Earth Data, such as multi-source remote sensing methodologies, can provide vital data support for assessing SDG 6.6.1.

In response to this need, the research team developed an innovative method for delineating the extent of natural lakes and reservoirs and created the Global Lakes (GLAKES) dataset. This dataset includes over 3.4 million natural lakes and reservoirs (with an area $\geq$0.03 km^2) and shows marked improvements over previous datasets in terms of temporal range, spatial coverage, and detailed delineation of small water bodies (Pi et al. 2022). This established a foundational resource for accurately and comprehensively capturing the spatiotemporal trends of natural lakes and reservoirs at the global scale. Utilizing GLAKES and various global datasets (Pekel et al. 2016; Donchyts et al. 2022; Wang et al. 2022), this case study assessed changes in the extent of natural lake and reservoir water bodies from 2001 to 2021 through remote sensing technology and Big Earth Data. By constructing a detailed surface water area time series dataset on a global scale, this case study explored the temporal trends, spatial variations, and underlying driving factors of the water area changes in global natural lakes and reservoirs, thereby providing critical support for water resource management and sustainable development.

- **Data**
  - The European Commission Joint Research Centre (JRC) Global Surface Water (GSW) dataset (Pekel et al. 2016) (https://global-surface-water.appspot.com/download).
  - GLAKES (Pi et al. 2022, https://zenodo.org/records/7016548).
  - The Georeferenced Global Dams and Reservoirs Dataset (GeoDAR) (Wang et al. 2022) and the global small- and medium-sized reservoirs dataset developed by Donchyts et al. (2022) (https://zenodo.org/records/6163413, https://figshare.com/articles/dataset/Supplementary_materials_for_the_High-resolution_surface_water_dynamics_in_Earth_s_small_and_medium-sized_reservoirs/20359860).

- **Method**

This case study first constructed a global water occurrence layer for each three-year period based on the monthly water classification layers from the JRC GSW dataset. Water occurrence refers to the proportion of observed instances classified as water to the total number of valid observations during a specific period, reflecting the frequency of water presence over the entire historical period. The water occurrence maps were then overlaid with the delineated water extent of GLAKES to construct a time-series dataset of water occurrence-weighted lake areas at different time periods. Based on this dataset, the temporal trends and spatial variations in natural lake and reservoir water area changes were analyzed at both $1° \times 1°$ grid and national scale, and their responses to both climate change and human activities were investigated for typical countries/regions.

The classification of natural lakes and reservoirs was determined by the spatial relationships between GLAKES and two global reservoir datasets. On the one hand, compared to the most comprehensive World Register of Dams (WRD), the total area and capacity of reservoirs recorded in GeoDAR accounted for 93.5% and 95.6% of those in WRD, respectively. On the other hand, the reservoir dataset constructed by Donchyts et al. included 71,208 small- and medium-sized reservoirs (0.1–100 km^2) globally, largely compensating for the gaps in GeoDAR regarding small- and medium-sized reservoirs. This case study combined both datasets to effectively characterize the changes in global reservoirs.

- **Results and Analysis**

1. **Spatial Distribution and Changes in Global Natural Lake and Reservoir Areas**

From 2001 to 2021, the average water area of global natural lakes and reservoirs reached 2.7×10^6 km^2, with reservoirs constituting 19.0% and natural lakes 81.0%. Figure 3.10 illustrates the proportion of water area in global reservoirs and natural lakes within each 1° × 1° grid, defined as the ratio of water surface area to grid area (range: 0%-100%). These water bodies can be further divided into three size groups: small (< 1 km^2), medium (1–100 km^2), and large (> 100 km^2). These size groups accounted for 95.53%, 4.43%, and 0.04% of the total number of water bodies, respectively, and 14.4%, 24.7%, and 60.9% of the total area, respectively.

From 2001 to 2021, the global coverage of natural lakes and reservoirs showed an overall expansion trend (Fig. 3.11). Among them, reservoirs exhibited continuous and significant expansion, while natural lakes underwent a cycle of initial shrinkage, subsequent expansion, and further shrinkage, leading to a slight overall decline. Specifically, the area change rate of reservoirs was 1,133.5 km^2/a, whereas natural lakes experienced a negative change rate of −414.4 km^2/a, primarily due to the considerable shrinkage in large natural lakes (−1,598.8 km^2/a). The growth rate of natural lakes in glacial and permafrost regions was 525.95 km^2/a, which was 2.72 times higher than that in non-glacial and permafrost regions, yet only 46.4% of the reservoir area expansion rate during the same period. This indicates that the overall water expansion from 2001 to 2021 was largely due to human-regulated reservoirs rather than replenishment from glacial melt attributable to climate change.

2. **Comparative Analysis of Natural Lake and Reservoir Areas in Different Regions**

Reservoirs are mainly concentrated in North America, Asia, and Europe, which together accounted for 81.4% of the total number and 70.6% of the total area globally. Developed countries represented 37.4% of the total number and 38.0% of the total area worldwide, while the corresponding proportions for the BRICS countries were 29.8% and 37.6%. In terms of natural lakes, North America and Asia had significantly higher area proportions compared to other continents, reaching 45.8% and 36.6%, respectively. Additionally, 51.0% of the number and 27.4% of the area of natural lakes were found to be located north of 60°N.

At the continental scale, except for South America, the global reservoir water area showed an increasing trend from 2001 to 2021, with Asia (835.4 km^2/a) and Africa (187.4 km^2/a) exhibiting significant expansion. At the national scale (Fig. 3.12a), the reservoir water area significantly increased in 46 countries, while only 7 countries showed a significant decreasing trend. The primary countries and regions displaying expansion trends include China, Russia, parts of Southeast and South Asia, Iran, Türkiye, Northeast Africa, Canada, and Greenland. Meanwhile, the countries and regions with shrinking trends are mainly concentrated in the United States, South America, Southern Africa, and the Northern Mediterranean coast.

Regarding natural lakes, no significant interannual area changes were observed from 2001 to 2021 on the continental scale, with positive changes only occurring in Africa, Europe, and North America. At the national scale (Fig. 3.12b), the lake water area significantly expanded in 68 countries, while 12 countries showed a significant shrinking trend. Countries or regions demonstrating expansion trends include China, parts of Southeast and South Asia, central and western Europe, the Arabian Peninsula, and the central

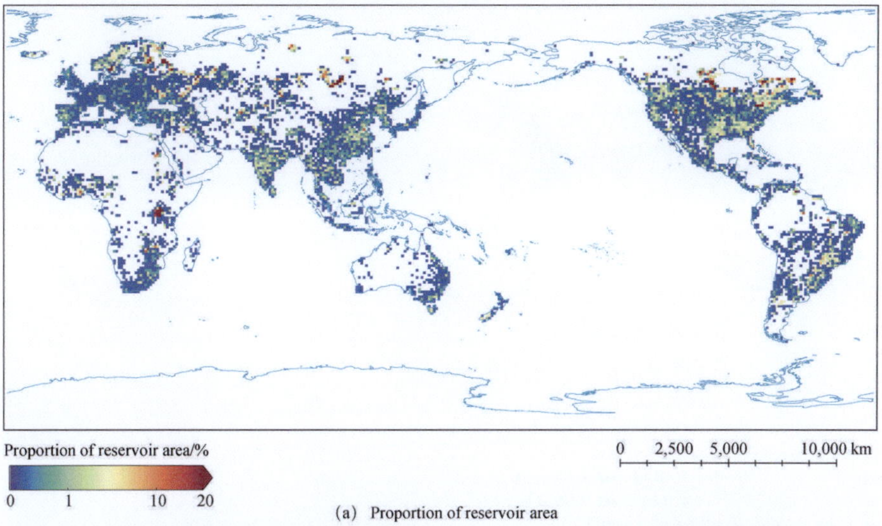

(a) Proportion of reservoir area

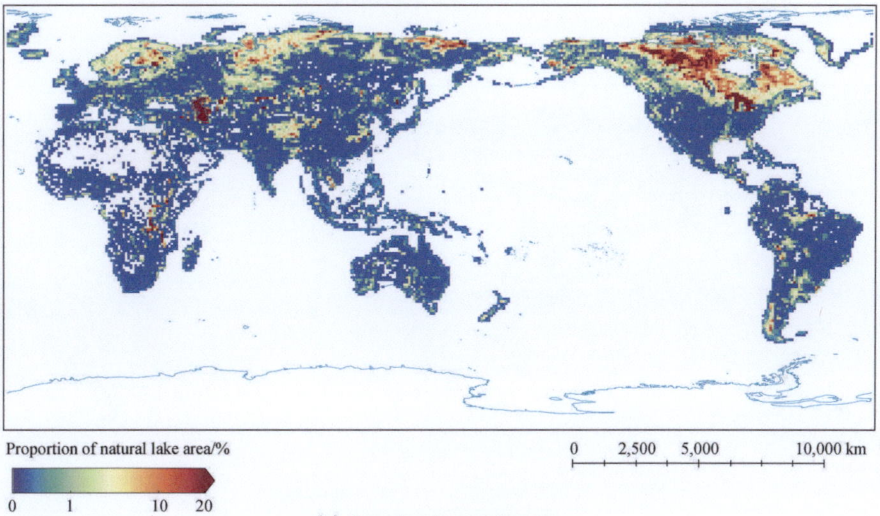

(b) Proportion of natural lake area

Fig. 3.10 Spatial distribution of the surface water area proportion of global reservoirs and natural lakes over the 2001–2021 period (1° × 1°)

and northwestern coast of Africa and Greenland. Meanwhile, countries or regions with shrinking trends are primarily located in West Asia, Ukraine, southern Africa, Australia, North America, and southern South America.

3. **Analysis of Changes in Natural Lake and Reservoir Areas in Typical Regions**

To further explain the characteristics of water area changes in global reservoirs and natural lakes, this case study selected 10 typical countries/regions worldwide based on factors including interannual water area change rates, geographical location, and the focus of similar studies. There were significant differences in the proportions of reservoir and natural lake areas among different countries/regions, with specific values shown in Table 3.2. Temporally, the areas of natural lakes and reservoirs in China, Sudan, and Greenland exhibited a significant expansion trend, while the reservoirs in the Democratic Republic of the Congo demonstrated a significant contraction trend, although natural lakes

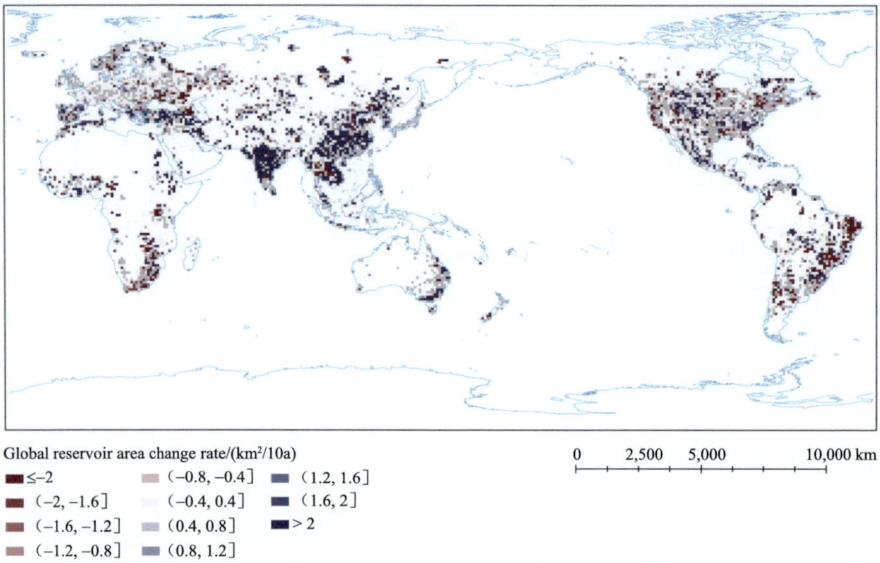

(a) Global reservoir area change rate

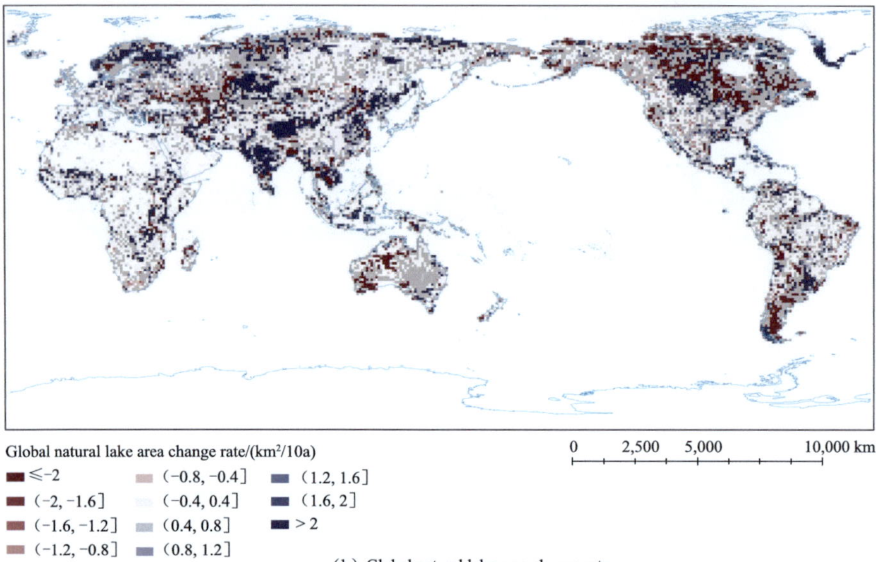

(b) Global natural lake area change rate

Fig. 3.11 Spatiotemporal pattern of the water area changes for global reservoirs and natural lakes over the 2001–2021 period (1° × 1°). *Note* The gray color indicates that the interannual change rate of the water area within this grid is not statistically significant

in this region showed a significant expansion trend. In addition, natural lakes in Kazakhstan were experiencing a significant declining trend, while reservoirs in Türkiye showed a significant expansion trend. No significant trends were observed in the water area changes in the remaining countries/regions.

A further investigation reveals that from 2001 to 2021, there were also variations in natural lake and reservoir areas within the inside regions of different typical countries/regions, with diverse causes for these differences (Fig. 3.13). In the northern plains of the United States, water bodies expanded due to a climatic progression from dry

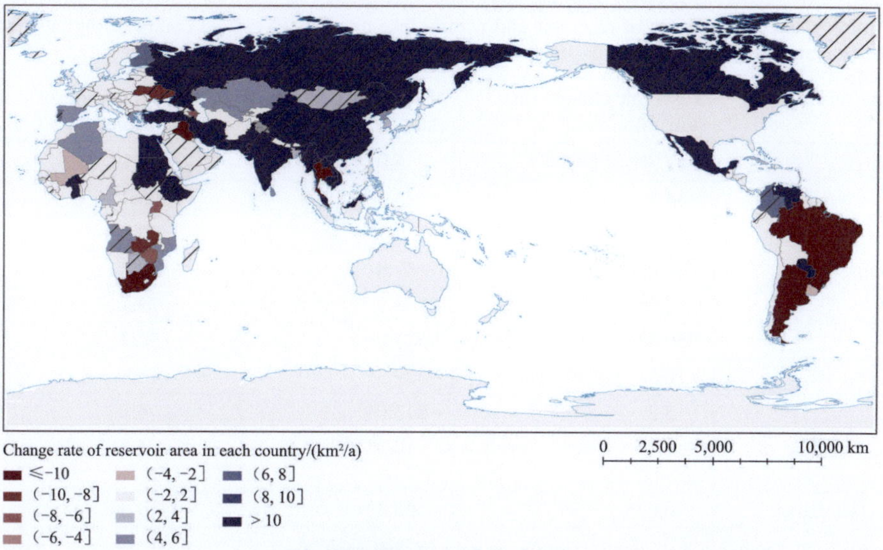

(a) Change rate of reservoir area in each country

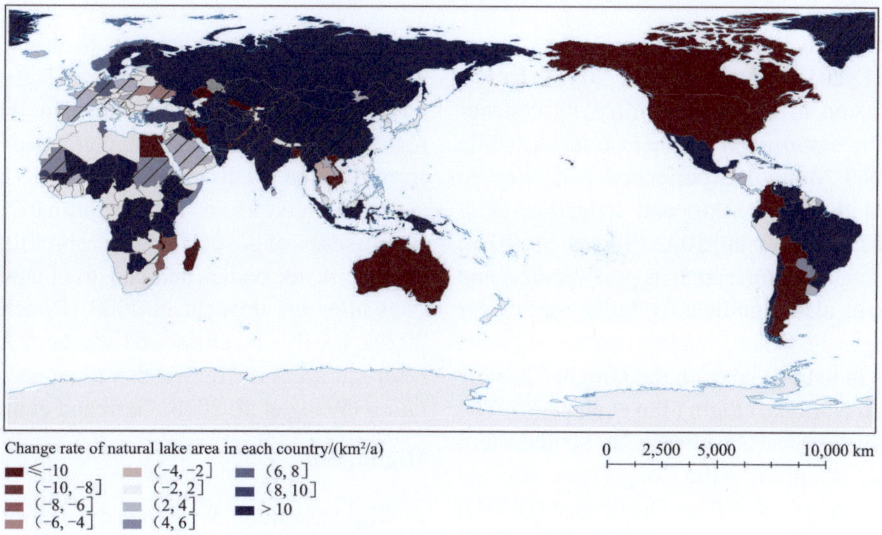

(b) Change rate of natural lake area in each country

Fig. 3.12 Interannual water area change rates of global reservoirs and natural lakes over the 2001–2021 period at the national scale. *Note* The hatched lines indicate that the interannual change rate of the water area in this country is statistically significant

to wet period, while those in the western coastal areas decreased due to drought and groundwater extraction (Scanlon et al. 2012; Rodell et al. 2018; Leeper et al. 2022). Natural lakes in Greenland (Denmark) expanded due to increased glacial melt resulting from climate warming (Shugar et al. 2020). In western Australia, natural lakes shrank due to a drying climate (Rodell et al. 2018), whereas natural lakes and reservoirs in the eastern regions expanded locally after experiencing the worst drought in over 100 years followed by recovery with heavy rains (van Dijk et al. 2013; Evans et al. 2012). The expansion of reservoirs in northern Sudan, as well as upstream natural lakes in the south, was associated with dam construction in the north and subsequent

Table 3.2 Interannual change rates of reservoir and natural lake areas in 10 typical countries/regions from 2001 to 2021 and the proportion of reservoirs in the total water area of each region

Typical countries/regions	Reservoir change rate/ (km^2/a)	Natural lake change rate/ (km^2/a)	Proportion of reservoir area/%
Sudan	67.3*	5.9*	93.1
Brazil	−34.4	23.5	45.4
Türkiye	40.9*	9.1	33.4
United States	-2.0	−47.8	24.2
Australia	2.0	−178.1	23.1
China	250.8*	577.0*	20.3
Argentina	−16.5	−329.3	16.8
Kazakhstan	5.2	−513.0*	4.2
Democratic Republic of Congo	−2.8*	10.0*	1.3
Greenland (Denmark)	0.02*	65.3*	0.06

Note An asterisk (*) denotes a statistically significant rate of change

upstream replenishment of water storage (Alrajoula et al. 2016; Ahmed et al. 2014). Reservoir expansion in China was primarily concentrated in the eastern and southern regions, while parts of Nei Mongol experienced a decline in water area due to mining and irrigation (Zhu et al. 2022; Song et al. 2022; Tao et al. 2020, 2015). Climatic factors, such as precipitation and glacial melt, also significantly influenced water area changes in China, leading to considerable expansion in natural lakes in the Qinghai-Xizang Plateau and Northeast Plain (Tao et al. 2020). The area dynamics of water bodies in the Southeastern Democratic Republic of the Congo were affected by the El Niño and Southern Oscillation (ENSO) and the Indian Ocean dipole (IOD) phenomena (Sogno et al. 2022). The contraction of the Caspian Sea and the Aral Sea largely contributed to the decline of natural lakes in the western region of Kazakhstan, primarily owing to local drought and excessive water extraction (Zmijewski and Becker 2014; Chen et al. 2017; Joodaki et al. 2014). In contrast, changes in natural lakes in the northeastern region were attributed to the melting of permafrost due to climate warming, which has led to the formation and expansion of numerous small thermal karst lakes (Polishchuk et al. 2015). In Central Brazil, the water area of natural lakes and reservoirs expanded due to drought alleviation and dam construction (Zarfl et al. 2015; Chen et al. 2010), while those in the eastern region declined because of continued suffering from drought (Getirana et al. 2016). The expansion of reservoirs in Türkiye primarily resulted from a series of government measures to replenish existing water bodies and construct new artificial ones after the drought in 2008 (Donchyts et al. 2022). Lastly, in central and southern Argentina, reservoir areas decreased due to reduced precipitation (Wang et al. 2018; Garreaud et al. 2017).

Highlights

- This case study developed the dataset detailing the global water occurrence every three years from 2001 to 2021.
- This case study investigated the temporal trends and spatial disparities in the area changes of global natural lakes and reservoirs across various scales.
- This case study analyzed the driving forces of the long-term area dynamics of natural lakes and reservoirs in selected countries/regions, providing decision-making support for achieving water-related sustainable development.

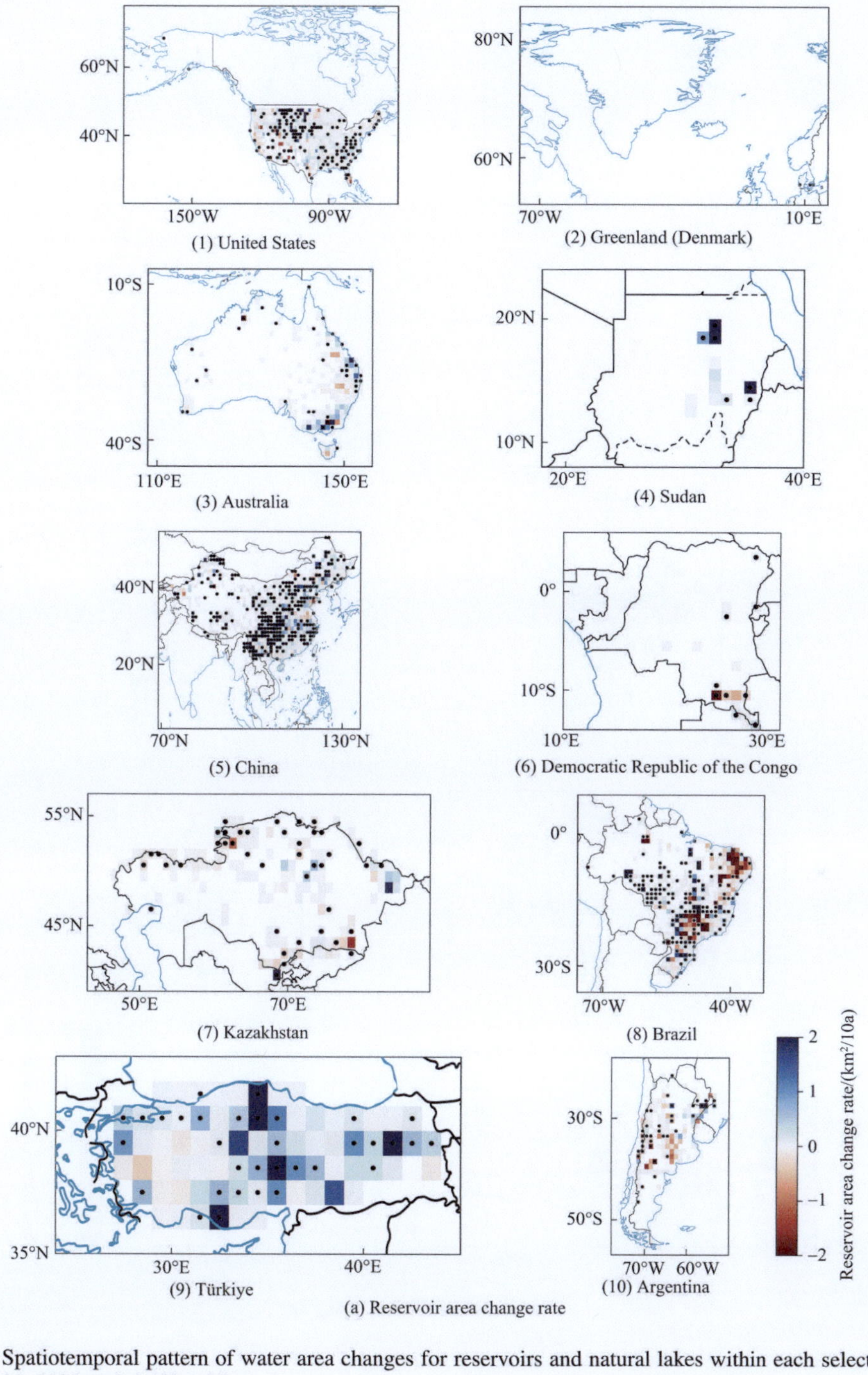

Fig. 3.13 Spatiotemporal pattern of water area changes for reservoirs and natural lakes within each selected country over the 2001–2021 period (1° × 1°)

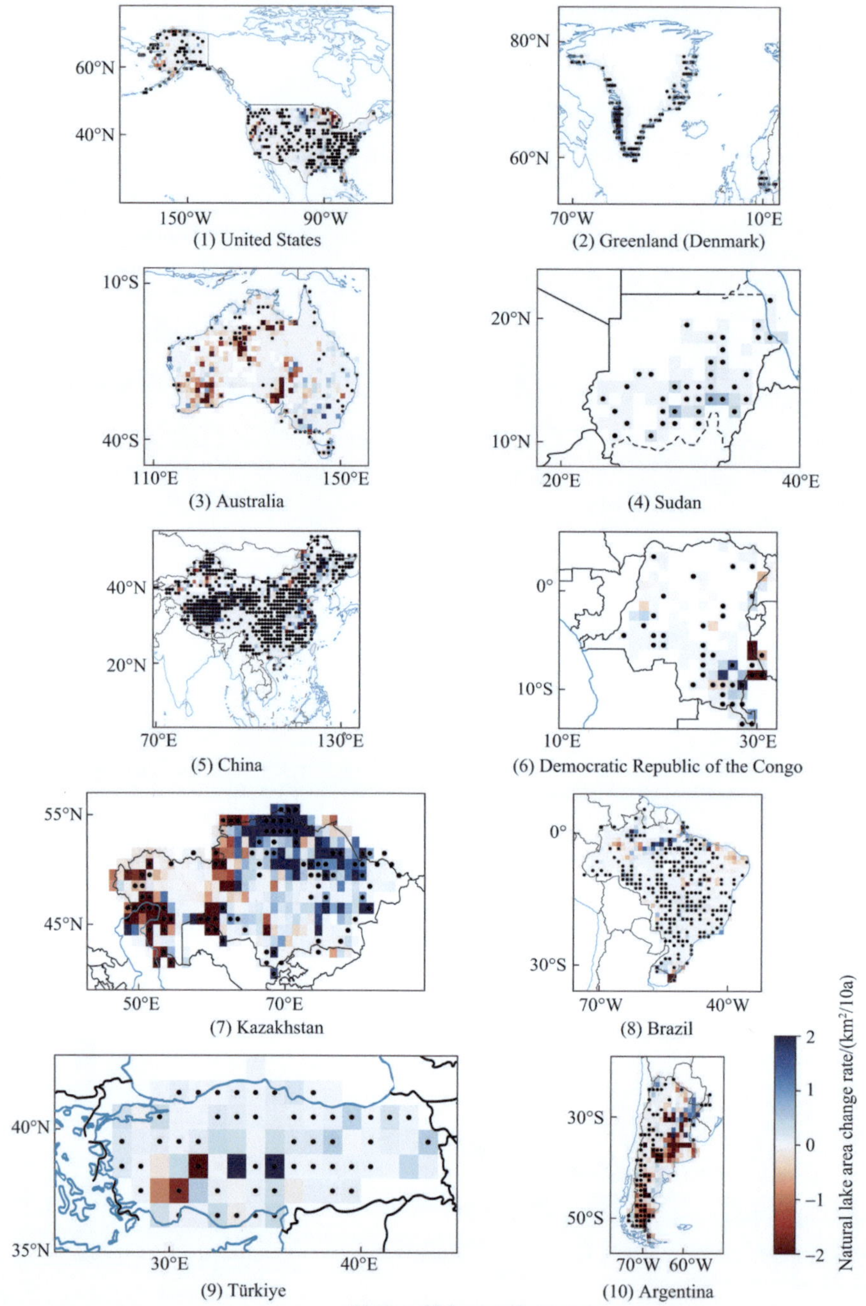

(b) Natural lake area change rate

Fig. 3.13 (continued)

- **Discussion and Outlook**

Leveraging remote sensing technology and Big Earth Data, this case study quantitatively evaluated changes in the water areas of global reservoirs and natural lakes between 2001 and 2021. It also examined spatial variations in water area changes across different countrie and regions, along with the mechanisms driving these changes. The findings indicate an overall expansion in the extent of global natural lakes and reservoirs during this period. At the continental scale, the water area of reservoirs increased in all continents except South America, with particularly notable expansion in Asia and Africa. In contrast, natural lakes did not exhibit significant interannual variation trends across the continents. Analysis of typical countries/regions identified glacial meltwater replenishment, increased precipitation, and reservoir water storage as key factors contributing to the expansion of natural lakes and reservoirs, whereas drought and excessive water extraction were associated with reductions in surface water area. The research findings can provide data support for countries to evaluate progress toward SDG 6.6.1, contributing to the protection and restoration of water ecosystems. To this end, countries/regions should closely monitor water resource status and take measures to protect freshwater resources, including drafting water resource management regulations, improving remote sensing monitoring networks, strengthening the construction of water resource storage and scheduling facilities, and raising public awareness of water resource protection through education. These measures are crucial for achieving the sustainable utilization of water resources.

3.4 Summary

This chapter, based on long-term global datasets of lake and reservoir water transparency, global cropland WUE dataset, and global water occurrence dataset derived from satellite remote sensing, provides an indirect assessment of changes in the SDG 6.3.2, SDG 6.4.1, and SDG 6.6.1 indicators from 2000 to 2021. The findings are as follows.

(1) The water transparency of global large lakes and reservoirs has shown an increasing trend. Noticeable improvements are observed in cold-region lakes and reservoirs, whereas low-transparency lakes in warm temperate zones exhibit a declining trend in transparency.
(2) Farmland WUE globally has been increasing, albeit with significant regional disparities. Advances in agricultural science and technology, substantial improvements in agricultural infrastructure, and the adoption of water-saving irrigation and agricultural water management techniques are the main drivers behind this improvement.
(3) Glacial meltwater replenishment, increased precipitation, and reservoir storage are the primary reasons for the expansion of water surface areas globally. Conversely, drought and over-extraction of water are the main causes of surface water area reductions in affected regions.

References

Ahmed M, Sultan M, Wahr J et al (2014) The use of GRACE data to monitor natural and anthropogenic induced variations in water availability across Africa. Earth Sci Rev 136:289–300
Ai ZP, Wang QX, Yang YH et al (2020) Variation of gross primary production, evapotranspiration and water use efficiency for global croplands. Agric for Meteorol 287:107935
Alrajoula MT, Al Zayed IS, Elagib NA et al (2016) Hydrological, socio-economic and reservoir alterations of Er Roseires Dam in Sudan. Sci Total Environ 566:938–948
Cai WY, Ullah S, Yan L et al (2021) Remote sensing of ecosystem water use efficiency: a review of direct and indirect estimation methods. Remote Sens 13(12):2393
Chen JL, Wilson CR, Tapley BD (2010) The 2009 exceptional Amazon flood and interannual terrestrial water storage change observed by GRACE. Water Resour Res 46(12):W12526
Chen JL, Pekker T, Wilson CR et al (2017) Long-term Caspian sea level change. Geophys Res Lett 44(13):6993–7001
Chen C, Park T, Wang XH et al (2019) China and India lead in greening of the world through land-use management. Nat Sustain 2:122–129
Cooley SW, Ryan JC, Smith LC (2021) Human alteration of global surface water storage variability. Nature 591(7848):78–81
Donchyts G, Winsemius H, Baart F et al (2022) High-resolution surface water dynamics in Earth's small and medium-sized reservoirs. Sci Rep 12(1):13776
Du DD, Zheng CL, Jia L et al (2022) Estimation of global cropland gross primary production from satellite observations by integrating water availability variable in light-use-efficiency model. Remote Sens 14(7):1722

Duan HT, Cao ZG, Shen M et al (2022) Review of lake remote sensing research. Natl Remote Sens Bull 26(1):3–18 (In Chinese)
Dubovyk O, Ghazaryan G, González J et al (2019) Drought hazard in Kazakhstan in 2000–2016: a remote sensing perspective. Environ Monit Assess 191(8):510
Evans JP, Boyer-Souchet I (2012) Local sea surface temperatures add to extreme precipitation in northeast Australia during La Niña. Geophys Res Lett 39(10):L10803
Garreaud RD, Alvarez-Garreton C, Barichivich J et al (2017) The 2010–2015 megadrought in central Chile: impacts on regional hydroclimate and vegetation. Hydrol Earth Syst Sci 21(12):6307–6327
Getirana A (2016) Extreme water deficit in Brazil detected from space. J Hydrometeorol 17(2):591–599
Hu GC, Jia L (2015) Monitoring of evapotranspiration in a semi-arid inland river basin by combining microwave and optical remote sensing observations. Remote Sens 7(3):3056–3087
Hu YF, Han YQ, Zhang YZ (2020) Land desertification and its influencing factors in Kazakhstan. J Arid Environ 180:104203
Joodaki G, Wahr J, Swenson S (2014) Estimating the human contribution to groundwater depletion in the Middle East, from GRACE data, land surface models, and well observations. Water Resour Res 50(3):2679–2692
Kang SZ (2022) Storing grain in water, storing water in technology—developing high water efficiency agriculture to ensure national food security. China Water Res 13:1–5 (In Chinese)
Karatayev M, Clarke M, Salnikov V et al (2022) Monitoring climate change, drought conditions and wheat production in Eurasia: the case study of Kazakhstan. Heliyon 8(1):e08660
Kavvada A, Metternicht G, Kerblat F et al (2020) Towards delivering on the sustainable development goals using earth observations. Remote Sens Environ 247:111930
Leeper RD, Bilotta R, Petersen B et al (2022) Characterizing U.S. drought over the past 20 years using the U.S. drought monitor. Int J Climatol 42(12):6616–6630
Ma RH, Yang GS, Duan HT et al (2011) China's lakes at present: number, area and spatial distribution. Sci China Earth Sci 54(2):283–289
Pekel JF, Cottam A, Gorelick N et al (2016) High-resolution mapping of global surface water and its long-term changes. Nature 540(7633):418–422
Pi XH, Luo QQ, Feng L et al (2022) Mapping global lake dynamics reveals the emerging roles of small lakes. Nat Commun 13(1):5777
Polishchuk YM, Bryksina NA, Polishchuk VY (2015) Remote analysis of changes in the number of small thermokarst lakes and their distribution with respect to their sizes in the cryolithozone of Western *Siberia*, 2015. Izv Atmos Ocean Phys 51(9):999–1006
Rodell M, Famiglietti JS, Wiese DN et al (2018) Emerging trends in global freshwater availability. Nature 557(7707):651–659
Scanlon BR, Faunt CC, Longuevergne L et al (2012) Groundwater depletion and sustainability of irrigation in the US high plains and central valley. Proc Natl Acad Sci USA 109(24):9320–9325
Shugar DH, Burr A, Haritashya UK et al (2020) Rapid worldwide growth of glacial lakes since 1990. Nat Clim Chang 10:939–945
Sogno P, Klein I, Kuenzer C (2022) Remote sensing of surface water dynamics in the context of global change: a review. Remote Sens 14(10):2475
Song CQ, Fan CY, Zhu JY et al (2022) A comprehensive geospatial database of nearly 100000 reservoirs in China. Earth Syst Sci Data 14(9):4017–4034
Spyrakos E, Hunter P, Simis S et al (2020) Moving towards global satellite based products for monitoring of inland and coastal waters. Regional examples from Europe and South America. In: 2020 IEEE Latin American GRSS & ISPRS remote sensing conference (LAGIRS). Santiago: IEEE, pp 363–368
Tao SL, Fang JY, Zhao X et al (2015) Rapid loss of lakes on the Mongolian Plateau. Proc Natl Acad Sci USA 112(7):2281–2286
Tao SL, Fang JY, Ma SH et al (2020) Changes in China's lakes: climate and human impacts. Natl Sci Rev 7(1):132–140
Tyler AN, Hunter PD, Spyrakos E et al (2016) Developments in Earth observation for the assessment and monitoring of inland, transitional, coastal and shelf-sea waters. Sci Total Environ 572:1307–1321
UNEP (2021b) Progress on integrated water resources management: tracking SDG 6 series: global indicator 6.5.1 Updates and Acceleration Needs. https://wedocs.unep.org/20.500.11822/36690
United Nations (2022) The sustainable development goals report 2022. https://desapublications.un.org/publications/sustainable-development-goals-report-2022
UN-Water (2018) Progress on ambient water quality: piloting the monitoring methodology and initial findings for SDG indicator 6.3.2. https://www.unwater.org/sites/default/files/app/uploads/2018/12/SDG6_Indicator_Report_632_Progress-on-Ambient-Water-Quality_ENGLISH_2018-1.pdf
UN-Water (2021) Summary progress update 2021: SDG 6-water and sanitation for all. https://www.unwater.org/sites/default/files/app/uploads/2021/12/SDG-6-Summary-Progress-Update-2021_Version-July-2021a.pdf
van Dijk AIJM, Beck HE, Crosbie RS et al (2013) The millennium drought in Southeast Australia (2001–2009): natural and human causes and implications for water resources, ecosystems, economy, and society. Water Resour Res 49(2):1040–1057
Wang JD, Song CQ, Reager JT et al (2018) Recent global decline in endorheic basin water storages. Nat Geosci 11(12):926–932
Wang SL, Li JS, Zhang B et al (2020) Changes of water clarity in large lakes and reservoirs across China observed from long-term MODIS. Remote Sens Environ 247:111949

Wang MH, Jiang LD, Mikelsons K et al (2021) Satellite-derived global chlorophyll-a anomaly products. Int J Appl Earth Obs Geoinf 97:102288

Wang JD, Walter BA, Yao FF et al (2022) GeoDAR: Georeferenced global dams and reservoirs dataset for bridging attributes and geolocations. Earth Syst Sci Data 14(4):1869–1899

Williamson CE, Saros JE, Vincent WF et al (2009) Lakes and reservoirs as sentinels, integrators, and regulators of climate change. Limnol Ocean 54(6part2):2273–2282

Woolway RI, Kraemer BM, Lenters JD et al (2020) Global lake responses to climate change. Nat Rev Earth Environ 1:388–403

Yang SS, Zhang JH, Wang JW et al (2022) Spatiotemporal variations of water productivity for cropland and driving factors over China during 2001–2015. Agric Water Manag 262:107328

Yu GR, Song X, Wang QF et al (2008) Water-use efficiency of forest ecosystems in Eastern China and its relations to climatic variables. New Phytol 177(4):927–937

Zarfl C, Lumsdon AE, Berlekamp J et al (2015) A global boom in hydropower dam construction. Aquat Sci 77(1):161–170

Zhai LC, Lü LH, Dong ZQ et al (2021) The water-saving potential of using micro-sprinkling irrigation for winter wheat production on the North China Plain. J Integr Agric 20(6):1687–1700

Zhang GQ, Yao TD, Chen WF et al (2019) Regional differences of lake evolution across China during 1960s–2015 and its natural and anthropogenic causes. Remote Sens Environ 221:386–404

Zhang YL, Qin BQ, Shi K et al (2020) Radiation dimming and decreasing water clarity fuel underwater darkening in lakes. Science Bulletin 65(19):1675–1684

Zhang B, Li JS, Shen Q et al (2021) Recent research progress on long time series and large scale optical remote sensing of inland water. Nat Remote Sens Bull 25(1):37–52 (In Chinese)

Zheng CL, Jia L, Hu GC et al (2019) Earth observations-based evapotranspiration in northeastern Thailand. Remote Sens 11(2):138

Zheng CL, Jia L, Hu GC (2022) Global land surface evapotranspiration monitoring by ETMonitor model driven by multi-source satellite earth observations. J Hydrol 613:128444

Zhu YN, Ye AZ, Zhang YH (2022) Changes of total and artificial water bodies in inland China over the past three decades. J Hydrol 613:128344

Zmijewski K, Becker R (2014) Estimating the effects of anthropogenic modification on water balance in the Aral Sea watershed using GRACE: 2003–12. Earth Interact 18(3):1–16

Zou J, Ding JL, Yang ST (2017) Spatial and temporal variation analysis of ecosystem water use efficiency in Central Asia and Xinjiang in recent 15 years. Geogr Res 36(9):1742–1754 (In Chinese)

4 SDG 7 Affordable and Clean Energy

4.1 Background

"Ensuring access to affordable, reliable and sustainable modern energy for all" is one of the global SDGs put forward by the UN 2030 Agenda. Reducing greenhouse gas emissions, curbing global temperature rise, and striving to realize the goal of carbon neutrality are the initiatives and common pursuits of mankind in the face of the climate change crisis.

The energy sector emits 3/4 of the world's greenhouse gases. And energy transition offers solutions to the climate crisis and energy security issues. Increasing the use of clean and low-carbon energy sources, especially wind and other renewable energy, is the key to the global energy transition and a crucial step towards achieving the Paris Agreement's 1.5 °C climate goal. SDG 7 sets out future goals for energy accessibility and transition, renewable energy, energy efficiency and international energy cooperation.

The development of renewable energy has become a global consensus. Technologies such as solar and wind have matured and become cost-competitive worldwide. According to the World Energy Transitions Outlook 2023: 1.5 °C Pathway issued by the International Renewable Energy Agency (IRENA), in order to achieve the goal of tripling the installed capacity of renewable energy by 2030, SDG 7 needs to be systematically advanced in planning, policy and regulations, and financing.

Despite the fact that the newly installed capacity of renewable energy has been hitting record highs in recent years, the energy transition is still insufficient. The parties to the Paris Agreement have added the topic of renewable energy to their nationally determined contributions (NDCs) targets, but only 148 countries have quantified the relevant targets. Since 2020, the frequency of global crises, the COVID-19 pandemic, climatic anomalies, and economic fluctuations has had a significant impact on energy security. The imbalanced energy development dynamics worldwide have resulted in a slow energy transition in many countries.

Strengthening international clean energy cooperation, with renewable energy as the main source, will enable developing countries to gain access to technology, financing, projects and human resources, and build a new win–win model for the green and low-carbon transformation of energy. China is a firm practitioner of the Paris Agreement, and its renewable energy technologies and industries, such as wind and solar power, are leading the world, providing a strong impetus for the world's energy transition. China's global intergovernmental energy cooperation mechanism has been established and improved, and its energy partnership has become increasingly close. In particular, China has made clean and low-carbon energy cooperation an important part of the joint construction of the

H. Guo, *Big Earth Data in Support of the Sustainable Development Goals (2023)—The Belt and Road*, Sustainable Development Goals Series, https://doi.org/10.1007/978-981-95-3178-3_4

"Belt and Road", and has carried out practical cooperation in the areas of energy investment, energy infrastructure connectivity, energy trade, and energy innovation, so as to join hands in addressing the common challenges of energy development and climate change.

This chapter provides policy insights and data support for decision-making by providing data products and methods and models to assess progress in advancing SDG 7 by utilizing Big Earth Data technology around SDG 7.

4.2 Main Contributions

This chapter assesses global and Chinese progress on the SDG 7.1, SDG 7.2, SDG 7.a, and SDG 7.b targets through 2 cases, with the main contributions shown in Table 4.1.

4.3 Case Study

4.3.1 The Role of International Cooperation Between China and Africa in Promoting African Solar Energy Utilization

Target: SDG 7.1 By 2030, ensure universal access to affordable, reliable and modern energy services.

SDG 7.2 By 2030, increase substantially the share of renewable energy in the global energy mix.

SDG 7.a By 2030, enhance international cooperation to facilitate access to clean energy research and technology, including renewable energy, energy efficiency and advanced and cleaner fossil-fuel technology, and promote investment in energy infrastructure and clean energy technology.

Table 4.1 Cases and their main contributions

Targets	Cases	Contributions
SDG 7.1 By 2030, ensure universal access to affordable, reliable and modern energy services	The Role of International Cooperation between China and Africa in Promoting African Solar Energy Utilization	Data product: Africa renewable energy dataset 2015–2022 Method and model: Methodology for assessing the impact of the China-Africa Energy International Cooperation Project on Africa's SDG 7 Decision support: Support for policy formulation and investment decisions on China-Africa energy cooperation
SDG 7.2 By 2030, increase substantially the share of renewable energy in the global energy mix		
SDG 7.a By 2030, enhance international cooperation to facilitate access to clean energy research and technology, including renewable energy, energy efficiency and advanced and cleaner fossil-fuel technology, and promote investment in energy infrastructure and clean energy technology	Evaluating Chinese Overseas Parks by Applying Low-carbon Development Indicator System	Data product: China overseas parks low-carbon development level dataset, in 2013 and 2019 Method and Model: Indicator system and criteria for low-carbon development in overseas parks Decision support: Providing guidance and demonstration for China's international energy cooperation
SDG 7.b By 2030, expand infrastructure and upgrade technology for supplying modern and sustainable energy services for all in developing countries, in particular least developed countries, small island developing States and landlocked developing countries, in accordance with their respective programmes of support		

- **Background**

China is the largest developing country in the world. The Chinese Foreign Minister has insisted on the first visit to Africa, the continent with the largest concentration of developing countries, at the beginning of each year for more than 30 consecutive years, which reflects the deep reality of cooperation and political foundation between China and Africa. Since the 1950s, China has been assisting Africa, providing maximum support to African countries in their struggle for national independence and economic development. Since the 1980s, with the principle of equality and mutual benefit, China-Africa relations have entered a new stage of comprehensive cooperation. In the new era, China adheres to the new concept of "truthfulness, realism, affinity and sincerity" in its cooperation with Africa, and has been strengthening cooperation with African countries under the "Belt and Road" cooperation to promote green development in Africa. Energy is the lifeblood of economic development, and solar energy is the most important renewable energy. China has become the world's largest country for solar energy production and installation for over 10 consecutive years, and has formulated a more complete policy system and industrial chain. The global share of solar photovoltaic production rose from 39% in 2008 to 78% in 2022. Africa is rich in solar energy resources, with more than 85% of the area having an annual solar radiation of 2,000 kW·h/m^2. According to the analysis of Statista, Africa has the largest potential for solar power generation in the world, with a long-term power generation capacity of more than 4.5 kW·h/kWp per day. The use of solar energy can help to alleviate the tense situation of electricity in African countries, achieve complementary advantages in China-Africa cooperation, and accelerate the realization of energy accessibility and transformation.

- **Data**

– Public database, World Bank (WB) (https://datatopics.worldbank.org/sdgatlas/goal-7-affordable-and-clean-energy?lang=en).
– Global statistical database (https://www.statista.com/chart/27722/practical-potential-solar-energy-output-by-global-region/).
– IRENA data (https://www.irena.org/How-we-work/Africa#key-data).
– Africa Solar Industry Association (AFSIA) (https://www.afsiasolar.com/data-center/reports/).
– Customs Statistics Online Query Platform, General Administration of Customs of the People's Republic of China (http://stats.customs.gov.cn/).

- **Method**

The solar energy sector is a priority point for China-Africa cooperation. China has been carrying out cooperation in investment and financing, product export, project construction, technical cooperation, personnel training, planning and livelihood improvement. Based on the improvement of indicators such as the electricity supply ratio, the total installed capacity of solar power, and the per capita installed capacity in Africa in 2022, the effectiveness of China-Africa cooperation in the solar energy field was assessed through a combination of qualitative and quantitative analytical methods.

- **Results and Analysis**

Under the global development and joint construction of the BRI, China has established a multi-level green cooperation system with Africa through South-South cooperation, the Forum on China-Africa Cooperation and other frameworks and mechanisms to jointly promote the development and utilization of solar energy in Africa. The GDI and the joint construction of the BRI have made green energy a key area of cooperation between China and Africa. China has organized a ministerial meeting of the Group of Friends of the Global Development Initiative, in which climate change and green development cooperation in 19 African countries were included in the first batch of projects. In the context of the BRI, a China-African Union energy partnership has been established. Relying on the FOCAC and other mechanisms, China and Africa have formulated and released policy documents such as the Program for China-Africa Cooperation in Economic and Social Development, the China-Africa Cooperation Vision 2035, the FOCAC Dakar Action Plan (2022–2024), and the Declaration on China-Africa Cooperation on Combating Climate Change (Zhang and Sun 2023). Green energy cooperation is a key focus of China's "Ten Cooperation Programs", "Eight Actions" and "Nine Projects" in support of Africa. According to the AFSIA, by the end of 2022, 30 countries had added more than 1 MW of solar energy, 16 countries had installed more than 10 MW, and 2 countries had added more than 100 MW. China-Africa cooperation projects on solar energy utilization cover the whole African continent (Fig. 4.1). Solar energy is the fastest-growing type of renewable

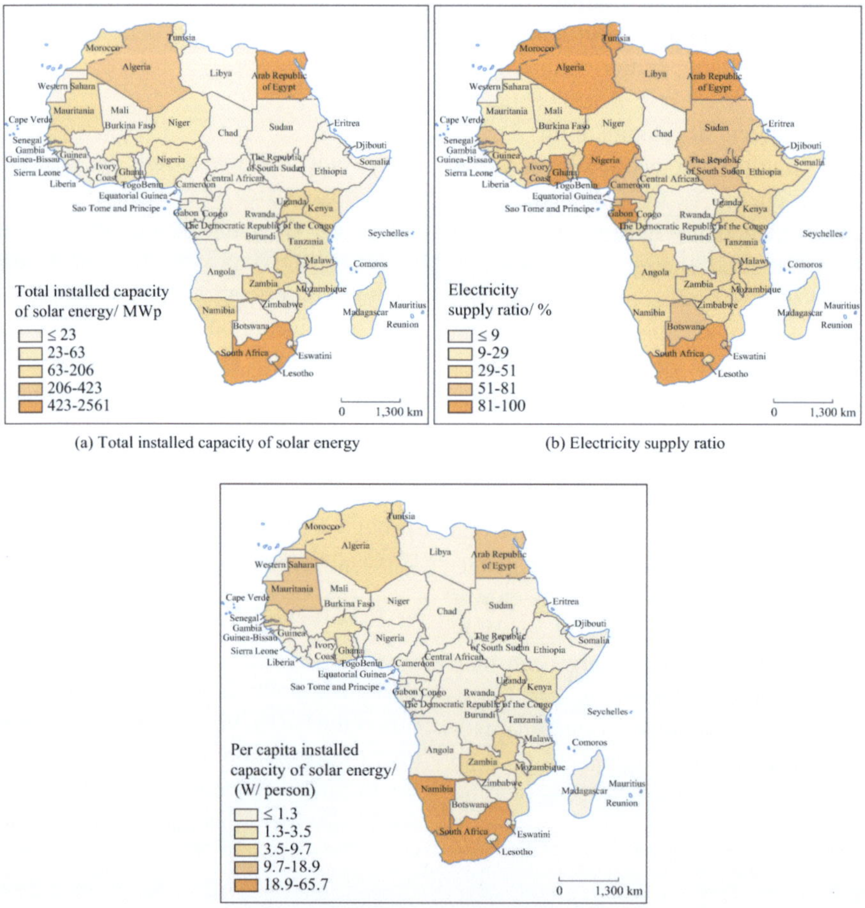

(a) Total installed capacity of solar energy

(b) Electricity supply ratio

(c) Per capita installed capacity of solar energy

Fig. 4.1 Total installed capacity of solar energy, electricity supply ratio, and per capita installed capacity of solar energy in African Countries in 2022 energy in Africa from 2015 to 2022, with the installed capacity increasing from 2,242 MW to 12,641 MW at an average annual growth rate of 28.03% (Figs. 4.2 and 4.3) (IRENA 2023a).

China-Africa international cooperation has effectively utilized Africa's abundant solar energy resources, effectively alleviated the problem of energy shortage, and enhanced energy accessibility in many African countries. IRENA statistics show that Africa's solar energy contains as much as 155,000 trillion-170,000 trillion Wh of electricity resources per year, but Africa's energy shortage is so serious. In 2010, the installed capacity of solar energy in sub-Saharan countries was only 40 MW, and over 600 million people in Africa lived in an environment without electricity. China's solar energy dominates the world, with a full industrial chain. Silicon wafers and PV module production have ranked first in the world for 12 and 15 consecutive years, accounting for 90% and more than 80% of the world respectively. China-Africa cooperation enables the joint development of African solar energy resources and further promotes solar energy utilization in Africa. The top 5 countries with installed capacity of PV power generation in 2022 included Angola (284 MW), South Africa (111.8 MW), Egypt (80 MW), Ghana (71.3 MW) and Mozambique (41.9 MW). From 2015 to 2022, China provided Africa with USD 5.531 billion of solar products (including 618,600 solar water heaters), with an average annual growth rate of 19.67% (Fig. 4.4). of this to tal, China exported 3.4 GW PV modules to Africa in 2022, a 36% year-on-year increase. China is the main provider of solar products in Africa, effectively supporting the development and utilization of solar energy resources in Africa.

China has actively cooperated with African countries to promote the utilization of solar energy through the provision of advice, financing support, project cooperation, product supply and capacity building, thus further promoting the development and utilization of solar energy in Africa and assisting Africa in realizing its energy transition. In 2021, Africa's investment in solar energy accounted for 65% of all renewable energy investments. The financial institutions in China have implemented Green Investment Principles for the BRI, have provided green development financing for Africa through the Silk Road Fund and other financial instruments, and have implemented hundreds of clean energy and green development cooperation projects. Chinese enterprises have built a number of solar projects through various investment and construction modes. For example, Chinese enterprises have invested in solar module factories in Cape Town; financed and constructed the largest 50 MW PV station in Garissa, Kenya, in the East African region; participated in the financing and construction of the 186 MW PV project of the Benban Photovoltaic Industrial Park in Egypt; financed, designed and built the Noor II and III solar thermal power plants in Morocco, which are the largest solar thermal power plants in the world; constructed the first PV power plant (15 MW) in Central African Republic; and supplied high-efficiency solar modules to the largest PV power station in Africa, the 233 MW PV plant in Algeria. In particular, China's traditional power plant infrastructure construction force in Africa has shifted to a green construction force after the country announced no new offshore coal power projects in 2021. 949 MW of new PV capacity was installed on the African continent in 2022, a 14% year-on-year increase compared to 833 MW in 2021. By 2021, the cumulative installed capacity of PV power plants built by Chinese enterprises in cooperation with Africa had exceeded 1.5 GW.

China has strengthened Africa's capacity for independent development and enhanced its capacity for sustainable development through scientific and technological cooperation, personnel training and planning. China's solar energy technology innovation leads the world, and China's PV industry has set the world record 14 times for crystalline silicon cell laboratory conversion efficiency in 2022. China has advanced technology transfer through the joint construction of technology platforms, and has enhanced the capability of solar energy utilization in Africa. For example, by relying on the China-Africa Environmental Cooperation Center, China promoted the "China-Africa Green Innovation Program"; China and Egypt jointly constructed the "China-Egypt National

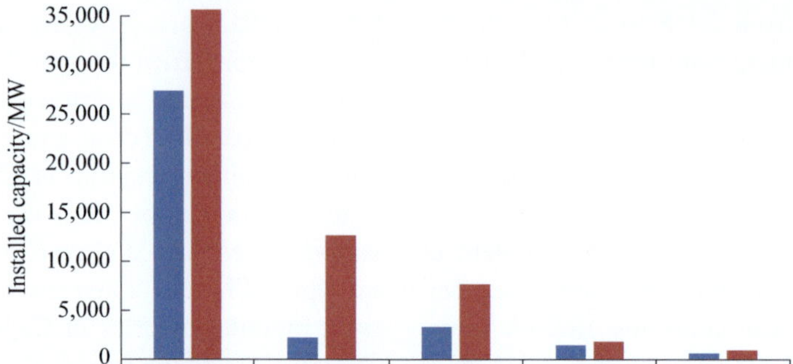

	Hydropower	Solar energy	Wind power	Biomass	Other
2015	27,343	2,242	3,320	1,440	626
2022	35,655	12,641	7,685	1,847	955

Fig. 4.2 Installed capacity of renewable energy by type in Africa in 2015 and 2022

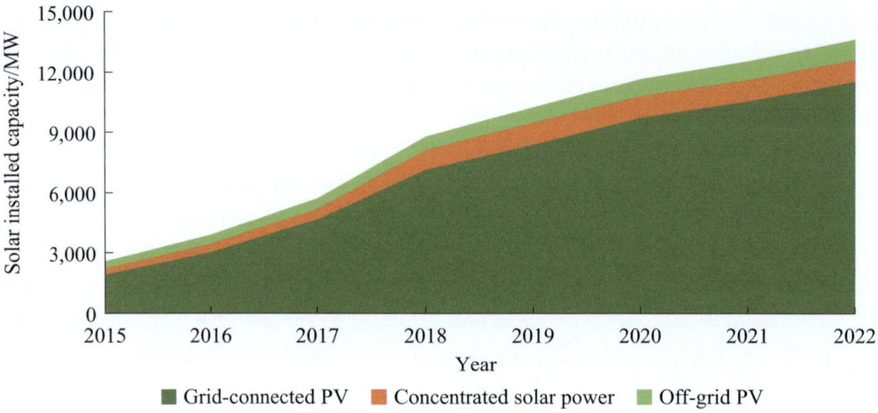

Fig. 4.3 Change in installed capacity of solar energy in Africa from 2015 to 2022

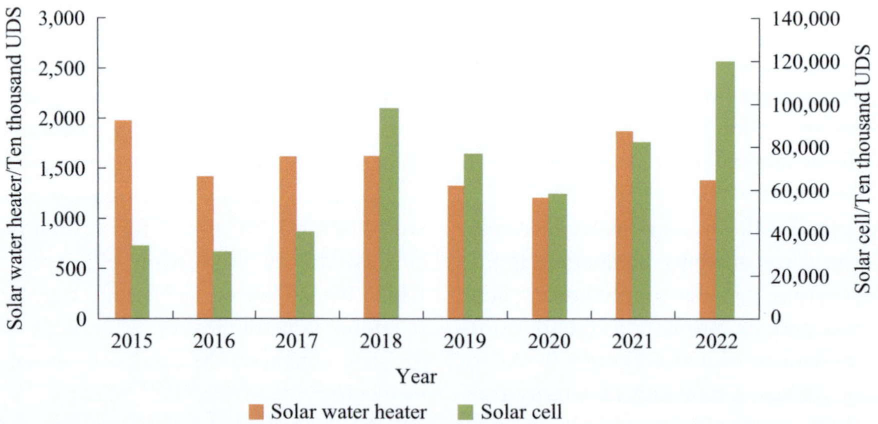

Fig. 4.4 China's exports of solar products to Africa (2015–2022)

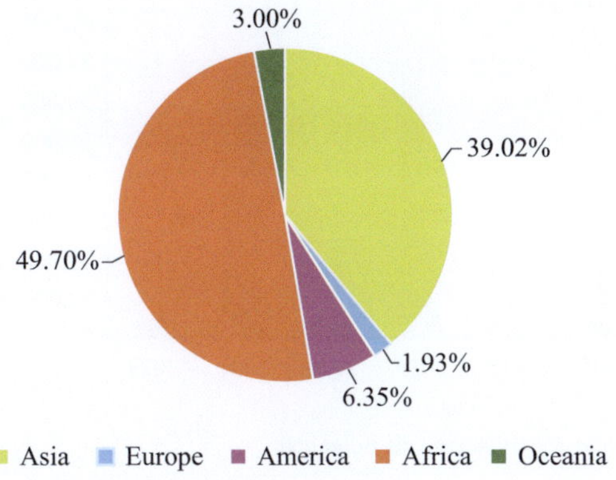

Fig. 4.5 Distribution of trainers for China's aid program on solar energy utilization (1991–2022)

Joint Laboratory on Renewable Energy" project; China set up the "China-South Africa Joint Research Center of Clean Energy" in South Africa; China and Kenya jointly implemented an international scientific and technological cooperation project on the utilization of solar energy; and Chinese enterprises provided "intelligent PV" to Africa and implemented "integrated photovoltaic and storage" new energy solutions in South Africa. In addition to providing solar products and technologies, China has also been training skilled personnel for Africa and enhancing their ability of solar energy utilization. For example, through the "Green Silk Road Envoy Program", China has trained green talents for Africa, and has opened solar energy courses in more than 10 "Luban Workshops" in Africa. In foreign aid training, the number of African trainers coming to China for solar energy utilization assistance accounted for about 50% of the total number (Fig. 4.5). China has utilized its own experience in energy transition and development to assist African countries in carrying out solar energy layout and planning. For example, China helped Ghana, Tanzania and Zambia to prepare plans for solar energy application, thereby upgrading Africa's capacity for solar energy development.

The "small but beautiful" projects in China-Africa solar energy cooperation focus on people's livelihoods, and contribute to the realization of SDGs by increasing jobs, and improving the education, healthcare and living environment of African residents. From 2010 to 2022, the global weighted average levelized cost of energy (LCOE) of PV declined from 0.445 USD/kW·h to 0.049 USD/kW·h, a decrease of 89%, which was already lower than the price of fossil energy (IRENA 2023b). This significant decrease is due in large part to the contribution of China, which has enabled more and more African countries to install and operate PV systems. According to Fig. 4.1, it can be seen that there were differences in the electricity supply ratio and per capita installed capacity of solar power across African countries in 2022. Although countries in the North Africa and Southern African regions do not have high rates of access to electricity, the amount of installed solar energy has effectively increased the watts per capita in most parts of Africa. The WB reports that more than 3,000 mini-grids had been installed in Sub-Saharan Africa by 2022, making it the largest market in the world. In industrial parks, mines, farms and remote areas where the grid is unstable or electricity costs are high, Chinese enterprises provide microgrid PV technology to realize power supply, and affordable household solar products to families in remote areas. Over 95% of the products certified by the WB's Lighting the World program are made in China. In terms of employment, it is estimated that the installation of each megawatt of solar PV system can create 12 jobs. Chinese enterprises build, operate and maintain PV power plants in Africa, employing local staff and boosting local employment. For example, Chinese enterprises provided about 5,000 local jobs during the construction of a PV power generation project in the Benban Photovoltaic Industrial Park in Egypt; and they also provided employment for about 700 people during the construction of the Sakai solar photovoltaic power plant in the Central African Republic in Central Africa. From 2015 to 2022, China donated 53,494 PV products to Africa and 950 sets of solar water heaters (Fig. 4.6). These products are widely used in water pumping, street lighting, power supply for telecommunication towers, rural schools, and clinics, effectively improving local healthcare, education, irrigation, and other infrastructure.

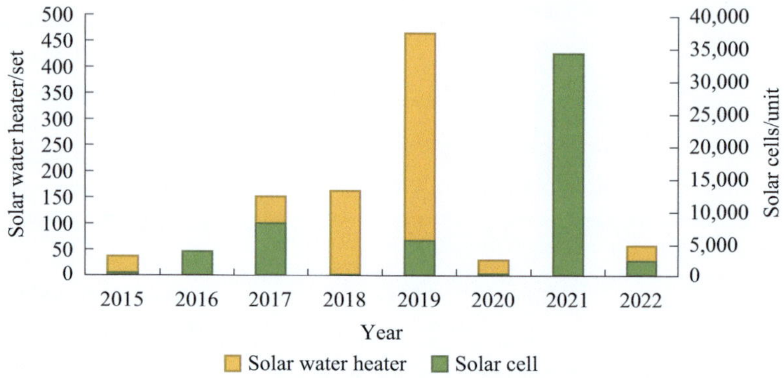

Fig. 4.6 China's donation of solar products to Africa (2015–2022)

Highlights

- From 2015 to 2022, the installed capacity of solar power in Africa grew from 2,242 MW to 12,641 MW, with an average annual growth rate of 28.03%, during which China supplied USD 5.531 billion solar products to Africa, with an average annual growth rate of 19.67%. By the end of 2022, 48 countries in Africa had possessed more than a megawatt of installed capacity of solar energy. The cumulative installed capacity of PV power plants built by China in Africa exceeds 1.5 GW, jointly contributing to the realization of SDG 7.1 and SDG 7.2.
- Through investment and financing, project construction, product supply, technology transfer, personnel training, and planning, the cooperation has enhanced Africa's access to electricity, has promoted improvements in employment, education, healthcare, and human settlements, and has contributed to the realization of SDG 7.a and other SDG indicators.

- **Discussion and Outlook**

(1) Africa's energy demand is increasing at an average annual growth rate of 6%, but the installed capacity of solar power accounted for only 0.5% of global additions in 2022, a proportion well below the world average. Sub-Saharan Africa, in particular, has a high demand for off-grid solar. Africa has the fastest growing population in the world, and its demand for energy will be even greater in the future. Meanwhile China ranks first in the world in terms of production capacity and output in the solar energy field. This provides a broad cooperation space for both sides to work together.

(2) From 2013 to 2022, about 60 million people in Africa had access to electricity, yet 72% of the rural population and 22% of the urban areas in sub-Saharan Africa still lack electricity. China is leading the global trend of cost reduction in solar energy and can provide affordable products to Africa. The concept of "small but beautiful" in particular injects new ideas into China-Africa solar energy cooperation and assists Africa in realizing energy accessibility as soon as possible.

(3) Africa is a key partner in China's BRI and an important part of South-South cooperation. Africa possesses rich solar energy resources and has great potential for development. China's solar industry is a leader in the world, with a complete industrial chain, and realized 100% nationwide electricity accessibility in 2015. China and Africa should take the opportunity of joint construction of the "Belt and Road" together to learn from China's development experience in the energy sector, strengthen solar energy cooperation, and jointly promote the realization of the global energy transformation.

4.3.2 Evaluating Chinese Overseas Parks by Applying Low-Carbon Development Indicator System

Target: SDG 7.a By 2030, enhance international cooperation to facilitate access to clean energy research and technology, including renewable energy, energy efficiency and advanced and cleaner fossil-fuel technology, and promote investment in energy infrastructure and clean energy technology.

SDG 7.b By 2030, expand infrastructure and upgrade technology for supplying modern and sustainable energy services for all in developing countries, in particular least developed countries, small island developing States and landlocked developing countries, in accordance with their respective programmes of support.

- **Background**

China's overseas parks play a critical role in boosting the development of China's overseas markets and the construction of global value chains, and reflect how the green BRI is conducted in the real world, considering that they tend to be energy-intensive areas with a greenhouse gas (GHG) emissions lock-in effect for a long service lifetime. Different from the "carbon peaking and carbon neutrality" practice of domestic parks in China, the low-carbon development of China's overseas parks is just starting up. Guidelines and indicator-system policies for low-carbon development targeting China's overseas parks have not yet been established. In addition, for the 201 China's overseas parks commonly mentioned in public or media reports, there is no research that systematically analyzes their low-carbon performance, which makes an incomplete picture of the state of play, and hinders the formulation and promulgation of relevant policies.

- **Data**

– Ministry of Commerce of the People's Republic of China, Report on Development of China's Outward Investment and Economic Cooperation (http://images.mofcom. gov.cn/fec/202102/20210202162924888.pdf).
– China Council for the Promotion of International Trade. Guide on the Business Environment of Foreign Countries for Chinese Investors (https://www.ccpit.org/yingshanghuanjing/).
– Belt and Road Portal (https://www.yidaiyilu.gov.cn/).

- **Method**

China's overseas parks are widely distributed geographically (Fig. 4.7), with different development stages, natural resource endowments and industrial categories, resulting in an uneven pattern of low-carbon development. Therefore, literature review and consultant interviews are conducted to summarize the existing low-carbon index system (Tian et al. 2018; Institute for Sustainable Communities 2012) for domestic parks and form on index pool with candidate indicators. We invited preeminent academics in this field to score the candidate indicators by holding seminars, and systematically identified 17 indicators to form the low-carbon development index (LCDI), covering: (1) low-carbon economy; (2) energy use; (3) resource use; (4) low-carbon facilities; and (5) low-carbon management (Fig. 4.8). Applying LCDI, this case study conducts a comparative assessment with 60 overseas parks, and aims to identify the development level and strengths the weaknesses of Chinese overseas parks in global parks by benchmarking global advanced parks.

- **Results and Analysis**

1. **The Level of Green and Low-Carbon Development Has Constantly Improved**

Comparing the LCDI of 2013 and 2019, the average score increased from 61.5 to 70.5, up by 14.6% (Fig. 4.9).

In terms of low-carbon economic development, the annual average output value growth rate

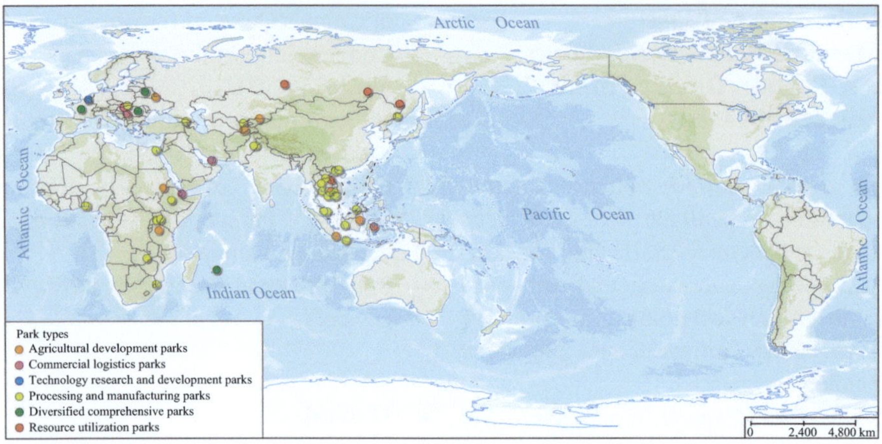

Fig. 4.7 Geographical distribution of 60 Chinese overseas parks

Criteria levels	Indicator levels
Low-carbon economy	• Unit construction land output • Proportion of low-carbon industries • Change rate of carbon emissions per unit output value • Relatively progressiveness low-caron development between overseas parks and local parks
Energy use	• Proportion of renewable energy consumption • Penetration rate of renewable energy • Energy consumption elasticity coefficient
Resource use	• Change rate of fresh water consumption per unit output value • Reuse rate of water in the park • Comprehensive utilization rate of solid waste
Low-carbon facilities	• Green coverage • Infrastructure completeness level • Proportion of green buildings • The completeness of green transportation facilities
Low-carbon management	• Low carbon management system/measures • Construction of energy/environmental management system • Low-carbon propaganda

Fig. 4.8 Low-carbon development indicator system of Chinese overseas parks

Fig. 4.9 Trend of LCDI in Chinese overseas parks. *Note* Among the 60 overseas parks, 40 were established before 2013, and 20 were established between 2014 and 2019. Therefore, only 40 overseas parks are listed in the figure

of overseas parks has maintained a high-speed growth trend. Some overseas parks have taken the initiative to undergo strategic transformation, continuously selecting green, low-carbon, and distinctive projects for cultivation, such as clean energy equipment manufacturing, waste incineration power generation, and seawater desalination. By 2019, the average proportion of low-carbon industries in overseas parks had reached about 13%, and the land utilization rate improved significantly compared to 2013, with an average unit construction land output of 0.148 million USD/m^2, which basically reached the advanced level of developed countries. The overseas parks have made extremely important contributions to promoting local economic development, driving local employment and tax growth, and advancing the industrialization process.

In terms of energy consumption and carbon emissions, the main types of energy consumption in verseas parks are electricity, raw coal, and natural gas. Due to the different leading industries of overseas parks, there are significant differences in the total energy consumption. The processing and manufacturing parks and resource utilization parks with high energy consumption industries such as steel, cement, petrochemicals, non-ferrous metals, building materials, and electricity have relatively high energy consumption, while the diversified comprehensive parks and commercial logistics parks with service industry have lower energy consumption. By 2019, the technology and capability enhancement measures taken by overseas parks and enterprises had significantly improved the technology and energy efficiency level of the host country, and the energy intensity and carbon emissions intensity of overseas parks had decreased year by year, with the average energy intensity at 1.21 tons of standard coal per USD 10,000, which was about 50% to 60% lower than the local average level and basically reached the advanced level of developed countries. At the same time, some overseas parks have actively carried out clean energy substitution and transformation to increase the proportion of renewable energy utilization. For example, the Ethiopia-Hunan Industrial Park uses 100% of the clean electricity provided by the largest wind power project in Africa—the Adama Wind Farm II in Ethiopia, and the Qilu (Cambodia) Special Economic Zone's biomass cogeneration project provides 80% of the park's electricity needs.

In terms of resource utilization, the measures for resource recycling in overseas parks mainly focus on reclaimed water reuse, comprehensive utilization of solid waste, and waste gas resource recovery, with 26.7%, 25.0%, and 10.0% of overseas parks taking relevant measures respectively. For example, the Liaoshen Industrial Park in Uganda recycles rainwater from sidewalks, roads, and building roofs for toilet flushing, road cleaning and green irrigation, significantly improving the efficiency of reclaimed water reuse within the park. The Djibouti International Free Trade Zone includes the generation, storage, utilization, and disposal of hazardous waste by enterprises entering the park in the supervision process, and provides solutions for toxic waste treatment for enterprises. The Uganda-China (Guangdong) Free Zone of International Industrial Park has achieved the recycling of wastewater, solid waste, waste heat, and various gases in the park through the linkage of upstream and downstream industrial chains. For example, tailings are used for brick making, and concentrates are used as fertilizer. However, we also need to recognize that most overseas parks did not consider the demand for resource recycling during the early planning, construction, and operation stages, resulting in low resource utilization efficiency. Therefore, it is urgent to accelerate the circular transformation of these parks.

In terms of low-carbon infrastructure, the performance of low-carbon infrastructure indicators is the best, with the highest average contribution rate to the LCDI of overseas parks. It is the core area where outstanding progress has been made in the low-carbon development of overseas parks. At present, average green coverage is around 30%, and the completion rates of centralized sewage treatment facilities and infrastructure are both around 60%. Some overseas parks have taken the lead in integrating green building and smart transportation concepts into park construction, creating a number of benchmark projects for overseas parks, and laying a solid foundation for the low-carbon transformation of the parks. For example, the SBIRD Innovation Base of the China Belgium Technology and Science Park has passed the building research establishment environmental assessment method (BREEAM) of the British Institute of Architecture and obtained a "very good" level certification. The Great Stone Industrial Park has promoted low-carbon travel and provided a public bicycle or charging station service system, laying a solid foundation for its energy transformation.

In terms of low-carbon operation and management, low-carbon operation and management in overseas parks are still weak areas, but some parks have carried out green and low-carbon management and practice and achieved initial results. Some parks have implemented green and low-carbon management and practices, such as setting up specialized departments to promote low-carbon work, establishing environmental management systems, and incorporating low-carbon concepts into their planning and design, thus ensuring the feasibility of low-carbon implementation from the source. In addition, some parks have provided the latest interpretation of environmental policies for enterprises and fully utilized expert resources in green and low-carbon to provide guidance for park enterprises, and have optimized the industrial structure by raising the entry threshold for the park and prioritizing the introduction of environmentally friendly enterprises.

2. **There Are Significant Differences in the Low-Carbon Development Level of Overseas Parks, Which Have Enormous Potential to Improve**

The national-level overseas parks are outstanding in terms of progressiveness and demonstration effect. Their low-carbon development level is 7.2% and 2.6% higher than provincial Overseas parks and other types of overseas parks (Fig. 4.10).

Regional differences and the economic development status of host countries significantly impact the low-carbon level of overseas parks. Overseas parks located in Europe (LCDI = 73.2) and Southeast Asia (LCDI = 68.7) have significantly higher levels of green and low-carbon development than those located in Central Asia, South Asia, West Asia, and Africa (Fig. 4.11).

The main industries in the processing and manufacturing parks and resource utilization parks are high-energy-consuming industries, such as steel, petrochemicals, mining and smelting, and non-ferrous metal smelting, with LCDIs ranging from 64.4 to 84.8. Their green and low-carbon development mainly focuses on two aspects: industrial low-carbonization and energy low-carbonization.

Processing and manufacturing parks with concentrated light industry have great potential for low-carbon development, but due to their low industrial interdependence and insufficient comprehensive utilization of energy and resources, their LCDIs range from 51.6 to 79.2. In the future, these parks need to promote coordinated development in industries, energy resources, infrastructure, and low-carbon management, and take multiple measures to promote low-carbon development in the parks.

The technology research and development parks have a solid industrial foundation and relatively low-carbon emission intensity. Their low-carbon development focuses on low-carbon energy, infrastructure, and low-carbon management. However, the space for reducing carbon emissions in agricultural development, commercial and logistics, and diversified comprehensive parks is limited. Therefore, energy and infrastructure low-carbonization are the focus of low-carbon development in these parks (Fig. 4.12).

Global advanced parks have achieved fruitful results in promoting green, low-carbon, and circular development, and are an important reference standard for the low-carbon development of China's overseas parks. By comparing with the benchmark values of advanced parks worldwide, it can be seen that the development trend of China's overseas parks is improving. However,

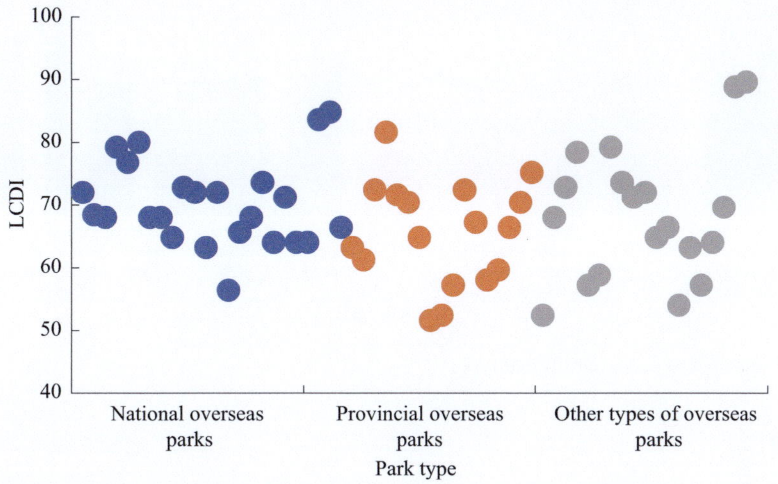

Fig. 4.10 Distribution for LCDI in various levels of overseas parks

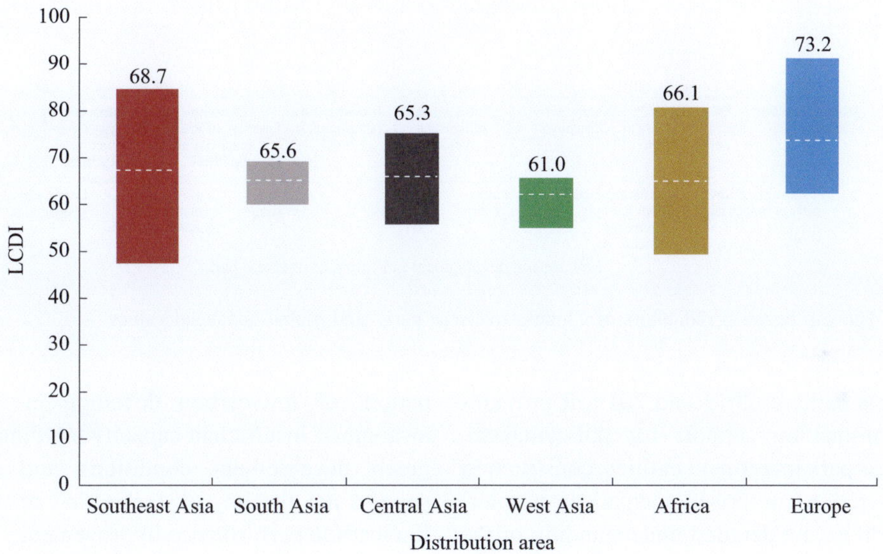

Fig. 4.11 LCDI of overseas parks in various regions. *Note* The white dashed line in the figure represents the average value of LCDI

in some areas, the low-carbon transformation efforts are limited, and the average value of 11 indicators is less than 50% of the benchmark value, which is a certain gap from the global advanced value. There is huge enhanced space for improvement in low-carbon development in the future (Fig. 4.13).

Highlights

- This case study provides the low-carbon development indicator system for China's overseas parks, constructs the LCDI for overseas parks, and compares and evaluates the low-carbon development level of 60 Chinese

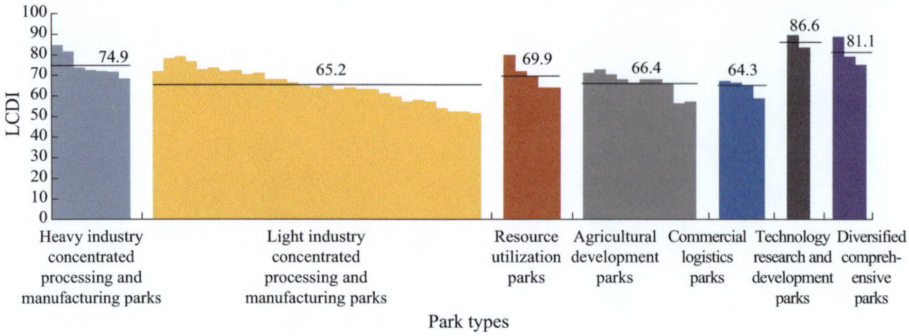

Fig. 4.12 LCDI of overseas parks of different types

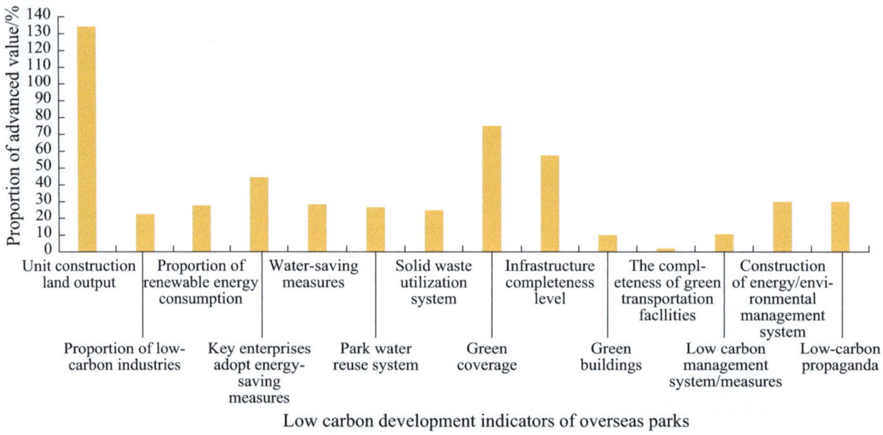

Fig. 4.13 The gap between indicators of Chinese overseas parks and global advanced values

overseas parks in 2013 and 2019. It provides decision-making support for policymakers, overseas park investment entities, construction and operation enterprises, and park entry enterprises to have a detailed understanding of the overall state, characteristics, and process of low-carbon development in China's overseas parks, as well as the formulation of relevant policies.

- **Discussion and Outlook**

In general, China's overseas parks are widely distributed geographically, with different development stages, natural resource endowments and industrial categories, resulting in an uneven pattern of low-carbon development. Lack of awareness, insufficient capacity building, limited green development conditions and financial support are the key obstacles that restrict their low-carbon performance improvement.

Firstly, overseas parks do not place enough emphasis on low-carbon development. Parks and enterprises mainly focus on economic growth and short-term profitability, lacking long-term planning for low-carbon development. China and the host country have not yet issued an overall plan and implementation rules for low-carbon development in parks. It is necessary to further improve the target management, guideline, progress evaluation, and assessment supervision of low-carbon development performance in overseas parks.

Secondly, overseas parks lack sufficient capacity building in low-carbon development. The understanding of concepts such as low-carbon, environmental protection, energy efficiency improvement, and circular economy among park managers, operators, and enterprises entering the park is still relatively vague. It is difficult to obtain and identify green, low-carbon technologies in key areas, and there are many obstacles to project implementation.

Thirdly, there is a lack of low-carbon development conditions in overseas parks. The energy resource infrastructure and transportation facilities in the area where the park is located are important factors determining the level of green and low-carbon development. At present, more than half of the overseas parks are located in areas with outdated power infrastructure; there are also many problems, such as unstable power supply, aging power equipment, and severe overload operation, which seriously affect the efficiency of equipment operation and energy utilization. At the same time, it also leads to the main reliance on fossil fuels for energy supply.

Fourthly, the insufficient scale of green and low-carbon investment and financing difficulty in overseas parks need to be improved. The incomplete diversification of green financial products and services makes it difficult for parks and enterprises to obtain financing from the capital market, and thus leads to a shortage of funds when implementing green and low-carbon projects in parks.

4.4 Summary

Focusing on the theme of international energy cooperation, this chapter summarizes China's experience in international energy cooperation and assesses the midterm progress of the global/China SDG 7 international cooperation indicators. It shows that since 2013, international cooperation has promoted global clean energy development and utilization in terms of energy infrastructure and resource development and utilization. Renewable energy, in particular, has become a reliable and affordable source of energy and will be the mainstay of future energy development. The following recommendations are made to advance the SDG 7 target.

(1) The development of wind, solar and other renewable energy sources is crucial to the realization of global energy transformation. Africa is abundant in solar energy resources, and China-Africa cooperation in the field of solar energy has effectively alleviated the problem of energy shortage in Africa, with the average annual growth of installed capacity in Africa reaching 28.03% from 2015 to 2022. In order to enhance the level of solar energy development and utilization in Africa and meet Africa's energy growth needs, China and Africa should continue to improve the utilization level of renewable energy in Africa under the frameworks and mechanisms of South-South Cooperation and the Forum on China-Africa Cooperation, through investment and financing, project construction, product supply, technology transfer, personnel training, planning and other modes of cooperation, and continue to assist the cost reduction of wind and solar PV in Africa, so that we can jointly achieve the goal of global energy transition.

(2) International energy cooperation is crucial for developing countries to achieve SDG 7. China has built more than 100 overseas parks in developing countries, which have made important contributions to promoting industrial upgrading, employment growth, revenue growth, and the construction of energy infrastructure in host countries. However, the pattern of green and low-carbon development in overseas parks is uneven due to the different development stages and energy structures of the host countries, outdated energy infrastructure, and inadequate green and low-carbon requirements and norms. In order to ensure the green attributes of investment activities, it is necessary to increase international cooperation in low-carbon scientific and technological innovation in the future, strengthen exchanges and cooperation in energy, environmental and climate governance policies

and standard systems, and thus provide guidance and demonstration for the construction of industrial chains and sustainable economic development in developing countries.

Big Earth Data has already played an important role in global energy accessibility and transition. In order to contribute to the realization of SDG 7 and China's "carbon peaking and carbon neutrality" goal, we will further explore the potential of Big Earth Data technology in the development of wind and solar resources, and provide scientific data and decision-making support services for global energy accessibility and transition.

References

China Council for the Promotion of International Trade (2022) Country (Region) Guide for Foreign Investment and Cooperation [R/OL]. https://www.ccpit.org/yingshanghuanjing/ [2024-04-19]

Institute for Sustainable Communities (2012) Guide for Low Carbon Industrial Development Zones [R/OL]. https://www.docin.com/p-737826912.html [2024-04-19]. (In Chinese)

IRENA (2023a) Renewable capacity statistics 2023 [EB/OL]. https://www.irena.org/-/media/Files/IRENA/Agency/Publication/2023/Mar/IRENA_RE_Capacity_Statistics_2023.pdf [2024-04-19].

IRENA (2023b) Renewable power generation costs in 2022 [EB/OL]. https://www.irena.org/-/media/Files/IRENA/Agency/Publication/2023/Aug/IRENA_Renewable_power_generation_costs_in_2022.pdf [2023-10-31]

Tian JP, Zang N, Xu Y et al (2018) Green development index of the Chinese national economic-technology development area [J]. Acta Ecol Sin 38(19):7082–7092 (In Chinese)

Zhang R, Sun TS (2023) China-Africa clean energy cooperation under global development initiatives [J]. Chinese J African Studies 4(1):62–82, 155–156. (In Chinese)

5 SDG 11 Sustainable Cities and Communities

5.1 Background

Achieving Sustainable Cities and Communities (SDG 11) is crucial for addressing climate change, promoting economic growth, and reducing poverty and pollution. By 2030, it is projected that 60% of the global population will reside in urban areas (Chen et al. 2022), and this proportion is expected to reach nearly 70% by 2050 (Sun et al. 2020). However, despite contributing 75% of the global GDP, urban areas also generate 75% of waste and carbon emissions worldwide. Rapid urbanization has led to challenges such as housing shortages, traffic congestion, air pollution, waste management issues, and inadequate infrastructure and services. Unplanned urban expansion makes cities more susceptible to the impacts of climate change and natural disasters.

SDG 11 aims to build inclusive, safe, resilient, and sustainable cities and human settlements, comprising seven technical targets and three policy targets. Among the 17 SDGs included in the UN 2030 Agenda, SDG 11 is interconnected with multiple other SDGs, and around one-third of the over 230 indicators can be measured at the city level. However, many indicators lack data support despite having evaluation methods, which poses a significant challenge in monitoring and assessing progress toward the targets and indicators. Big Earth Data technology, such as satellite remote sensing and geographic information system (GIS), can regularly obtain global data, uncover drivers of progress toward SDG 11, and serve as important data sources and analysis tools for global sustainable development research.

The 2019–2022 reports monitored and evaluated the progress of SDG 11 globally and in China, providing strong support and guidance for urban sustainable development. With a focus on targets, such as urban public transport (SDG 11.2), urban land use efficiency (SDG 11.3), heritage preservation (SDG 11.4), urban disaster management (SDG 11.5), urban air pollution (SDG 11.6), and open public spaces in cities (SDG 11.7), this report assesses the proportion of the global population with convenient access to public transport, develops a global urban boundary spatial distribution dataset, supplements boundary data on World Natural Heritage Sites and Mixed Cultural-Natural Heritage Sites, evaluate global extreme climate events from 2000 to 2022, analyzes and monitor changes in global atmospheric particulate matter concentrations, and examines changes in urban greenness and beneficiary populations in major cities worldwide. This chapter will focus on the progress assessment of SDG 11 at the global scale, providing a scientific foundation and experiential reference to reveal the current status of the global implementation of SDG 11 and to support the achievement of SDG 11.

H. Guo, *Big Earth Data in Support of the Sustainable Development Goals (2023)—The Belt and Road*, Sustainable Development Goals Series, https://doi.org/10.1007/978-981-95-3178-3_5

5.2 Main Contributions

This chapter evaluates six specific global targets—SDG 11.2, SDG 11.3, SDG 11.4, SDG 11.5, SDG 11.6, and SDG 11.7 through six case studies, with the main contributions shown in Table 5.1.

5.3 Case Study

5.3.1 Proportion of Population in Major Global Cities with Convenient Access to Public Transportation and Its Changes

Target: SDG 11.2 By 2030, provide access to safe, affordable, accessible and sustainable transport systems for all, improving road safety, notably by expanding public transport, with special attention to the needs of those in vulnerable situations, women, children, persons with disabilities and older persons.

- **Background**

According to a survey by the UN, more than 50% of the world's population currently resides in urban areas, which is expected to exceed 70% by 2050. However, the rapid growth of urban populations has put immense pressure on urban planning, transportation, the environment, and public health. The world's major cities occupy only 3% of the Earth's land area, producing 75% of the world's carbon emissions (Takase 2018). The acceleration of urbanization will lead to an increase in individual residents' demand for transportation. To alleviate traffic pressure and reduce carbon emissions, shifting travel demand to public transportation is feasible (Mittal et al. 2016). Efficient and reliable urban public transportation systems can actively promote the sustainability of transportation development, reduce urban congestion and environmental problems, and promote socio-economic growth and social progress (Saif et al. 2018). Therefore, achieving sustainable urban public transportation is also one of the very important areas in the SDGs.

Research on the accessibility of urban public transportation is still incomplete. Some studies mainly focus on solving the accessibility problem of public transportation systems in small cities (Tiwari and Phillip 2021), insufficiently consider the connectivity between different public transportation modes (Saghapour et al. 2016) or only study a single bus system (Chen et al. 2019), and lack a systematic evaluation of the spatiotemporal evolution at the global scale. Therefore, this case study aims to reveal the proportion of the population that can conveniently use public transportation and its changes in major cities worldwide based on multimodal spatiotemporal big data.

To achieve SDG 11.2, we must be people-oriented, focus on achieving the accessibility and quality of public transportation, adopt scientific and sustainable technologies and strategies, promote the development of urban public transportation, and improve residents' transportation experience, to enable residents to have convenient, low-carbon and high-quality travel options in the city.

- **Data**

– 2015 and 2020 global public transportation [including bus, bus rapid transit (BRT), light rail, subway, and ferry] station point of interest (POI) data obtained from OpenStreetMap (OSM).
– 2015 and 2020 global population grid WorldPop data, with a spatial resolution of 100 m.
– 2019 Global Human Settlement-Urban Centre Database (GHSL-UCDB) data, with a spatial resolution of 1 km.

- **Method**

Based on GHSL-UCDB and WorldPop data, 330 major cities with a population of more than 1 million were extracted through overlay analysis. Based on OSM POI data, public transportation station data of the above cities were extracted, and according to the UN recommendations, a buffer

Table 5.1 Cases and their main contributions

Targets	Tiers	Cases	Contributions
SDG 11.2 By 2030, provide access to safe, affordable, accessible and sustainable transport systems for all, improving road safety, notably by expanding public transport, with special attention to the needs of those in vulnerable situations, women, children, persons with disabilities and older persons	Tier I	Proportion of Population in Major Global Cities with Convenient Access to Public Transportation and Its Changes	Data product: Proportion of population with convenient access to public transport in 303 major global cities and its changes in 2015 and 2020 Method and model: Public convenient accessibility to public transport (PCAPT) and method for calculating population proportion with convenient access to public transport Decision support: Bus stop optimization recommendations
SDG 11.3 By 2030, enhance inclusive and sustainable urbanization and capacity for participatory, integrated and sustainable human settlement planning and management in all countries	Tier I	Global Urban Boundary Spatial Distribution Product	Data product: Global urban boundary spatial distribution dataset with multiple time phases from 2000 to 2022 Method and model: Using a combined approach of kernel density estimation (KDE) and urban cellular automata models to fill internal non-urban areas for initial urban boundary delineation, employing morphological methods to extract detailed urban boundaries, and analyzing spatiotemporal characteristics of urban area changes among global urban clusters Decision support: Provide a basic spatial unit for urban planning, urban development evaluation and other applications, supporting research on the impact of urbanization and other human socio-economic activities on surrounding ecology and environment
SDG 11.4 Strengthen efforts to protect and safeguard the world's cultural and natural heritage	Tier II	Update and Improvement of Global Public Boundary Data for World Natural Heritage Sites and Mixed Cultural-Natural Heritage Sites	Data product: Updated vector boundary data for 257 World Natural Heritage Sites and Mixed Cultural-Natural Heritage Sites Method and model: Heritage area data variance rate Decision support: Support WHC, IUCN and HIST World Heritage Data Development Initiative, contributing to heritage preservation and monitoring
SDG 11.5 By 2030, significantly reduce the number of deaths and the number of people affected and substantially decrease the direct economic losses relative to global gross domestic product caused by disasters, including water-related disasters, with a focus on protecting the poor and people in vulnerable situations	Tier I	Global Assessment Report on Extreme Weather and Climate Events from 2000 to 2022	Data product: Global extreme climate event dataset 2000–2022 Method and model: Objective disaster event identification methods, multi-source data fusion methods Decision support: By integrating data from emergency disaster databases and global disaster data platforms, a dataset of global extreme climate events from 2000 to 2022 has been created, addressing some of the data gaps in SDG 11.5.1 and SDG 11.5.2

(continued)

Table 5.1 (continued)

Targets	Tiers	Cases	Contributions
SDG 11.7 By 2030, provide universal access to safe, inclusive and accessible, green and public spaces, in particular for women and children, older persons and persons with disabilities	Tier I	Global Change in Urban Greenness and Beneficiary Population in Large Cities	Data product: Global 250 m spatial resolution greenness trend dataset 2001–2022 Method and model: Using non-parametric estimation models to calculate the greenness trend and significance level for each city, then determining the urban significant greening ratio by calculating the proportion of significantly greened urban area to the total urban built-up area (UBA) Decision support: Provide low-cost solutions for monitoring urban sustainable development in developing countries
SDG 11.6 By 2030, reduce the adverse per capita environmental impact of cities, including by paying special attention to air quality and municipal and other waste management	Tier I	Spatiotemporal Trends of Atmospheric Particulate Matter in Global Typical Regions	Data product: Global atmospheric particulate matter station monitoring dataset Method and model: Data fusion methods based on multiple data quality control methods Decision support: Reveal the current state of global air pollution, helping environmental departments in different regions develop locally appropriate environmental management policies

zone with a straight-line distance of 500 m was established for low-capacity public transportation stations (including buses and BRT), and a buffer zone with a straight-line distance of 1 km was established for high-capacity public transportation stations (including light rail, subway, and ferry). Each city's buffer zone, city center, and population grid data were superimposed to calculate the proportion of the population that can conveniently use public transportation, that is, the total population falling into the buffer zone divided by the city's total population. The SDGs aim to thoroughly solve the development problems in the three dimensions of society, economy, and environment through a comprehensive approach from 2015 to 2030 and promote the transition to sustainable development. By analyzing the data from 2015 and 2020, the progress of SDG indicators within 5 years was proposed, and further optimization and improvement directions were established. Comparing the data from 2015 and 2020 can reveal the changing trend of urban public transportation convenience over time, compare the differences between different regions and periods, and measure the degree of inequality. This method calculated the proportion of the population that can conveniently use public transportation in different regions (globally and in six continents) and different periods (2015 and 2020) (weighted by urban population) and introduced the Gini coefficient to analyze the imbalance of convenient public transportation in major cities around the world at different times.

The Gini coefficient is a commonly used indicator to measure the degree of uneven distribution. Its value ranges from 0 to 1, with 0 indicating complete equality and 1 indicating complete inequality. The Gini coefficient has the advantages of being easy to understand, having a multi-scale sample comparison, and cross-temporal and spatial analysis. It can better show the differences between data. Meanwhile, it can reflect the degree of difference in public transportation accessibility between cities, regions, and periods. After introducing population parameters, a horizontal comparison of the unevenness of transportation accessibility in 330 cities worldwide was conducted. This analysis shows the distribution pattern of public transportation resources among urban populations and effectively reveals the social fairness of urban public transportation accessibility. However, the Gini coefficient also has some limitations. It cannot capture the specific factors behind the allocation of transportation resources and is also limited by data quality and scope.

The proportion of the population that can conveniently use public transportation refers to the buffer zone established with the distance of convenient use of public commuting facilities as the radius. The population of the buffer zone within different urban divisions was extracted using the partition extraction method. The population within the buffer zone was compared with the city's total population. The obtained ratio coefficient is the proportion of the population that can conveniently use public transportation. By calculating the proportion of the population that can conveniently use public transportation for daily commuting, we can reflect the development level and service efficiency of urban public transportation.

$$R = \frac{R_s}{R_r}$$

where, R is the proportion of the population that can conveniently use public transportation, R_s is the population that can conveniently use public transportation within the buffer zone, and R_r is the total population of the city.

The Gini coefficient was first proposed by Italian statistician Corrado Gini in 1912. It is a quantitative indicator based on the Lorenz curve and represents the balance of resource distribution. Since the Gini coefficient calculation methods may differ in different countries, this case study adopts the Gini coefficient calculation method proposed by Tian (2012). The Gini coefficient reflects the imbalance. The larger its value, the greater the difference in the proportion of the population that can conveniently use public transportation between different cities. The calculation formula for the Gini coefficient is as follows:

$$G = 1\frac{1}{R \times \mathcal{W}} \sum_{n=1}^{n} (W_{i-1} + W_i) \times P_i$$

where, G is the Gini coefficient, P is the proportion of the population that can conveniently use public transportation, $\mathcal{W}$ is the total population of the city, i is the city serial number, n is the total number of typical cities in the experiment, W_i and P_i represent the total population of cities accumulated up to the i group and the corresponding accumulated proportion of the population that can conveniently use public transportation, respectively.

- **Results and Analysis**

In 2020, the weighted average of the population with convenient access to public transportation in major cities worldwide exceeded 60%. Among them, the values in cities in Europe and East Asia are relatively high (> 80%), while the values in Africa and South Asia are relatively low (<20%) (Fig. 5.1). It can be seen that there are significant differences in the convenience of public transportation in major cities around the world.

The proportion of people with convenient access to public transportation in major cities worldwide has increased significantly, but it is still uneven. The population-weighted mean of the world's major cities increased by 3.38% in 2020 compared with 2015; about 1/7 of the world's cities increased by more than 10%, mainly in regions outside Europe. In 2015, cities with a low proportion of the population with convenient access to public transportation showed significant improvement. The Gini coefficient of the PCAPT indicator in 330 major cities worldwide in 2020 was 0.59, basically the same as in 2015 (if the value is less than 0.4, it is considered relatively balanced). Judging from the changes in PCAPT values, regions with higher values in 2020 have relatively smaller changes. However, the numerical changes in the Middle East are more significant than in other regions; regions with higher values in Africa in 2020 have larger value increases. However, the changes in regions with lower values from 2015 to 2020 are not obvious, indicating that the development gap between regions with higher and lower PCAPT values in Africa has widened recently (Fig. 5.2).

Combined with the distribution of the top 300 cities in the world by GDP in Fig. 5.3, cities with higher PCAPT indices are mainly distributed in economically developed regions. The index values in Europe and East Asia are much higher than the global average. At the same time, those in Southeast Asia, the Middle East, and Africa are generally lower than the global average. For the Americas, cities with prosperous economies and politics have higher PCAPT indices, while others have relatively low PCAPT indices. There are obvious differences in the development level of public transportation worldwide. Economic prosperity can promote investment in public transportation infrastructure and improve transportation convenience and connectivity. There is a close connection between economic development and the convenience of public transportation.

In 2020, the population-weighted mean of the proportion of the population that can conveniently use public transportation in European cities was about 2.3 times that of African cities. Compared with 2015, the population-weighted mean of major cities in each continent showed an upward trend. The change order shows a pattern of South America (10%) > Africa (7%) > North America (6%) > Oceania (5%) > Asia (2%) > Europe (0%) (Fig. 5.4), indicating that cities with a low proportion of the population that can conveniently use public transportation in the previous period (2015) have significantly improved.

In 2020, the population-weighted mean of the PCAPT indices in major cities in developed countries was about 1.9 times that of the least developed countries (Fig. 5.5). Compared with 2015, the population-weighted mean of the PCAPT indices in major cities in developed countries, developing countries, and least developed countries showed an upward trend, and the upward trend in the least developed countries was the most obvious, about 9%. Even so, among the 330 major cities in this case study, nearly 400 million people in developing countries still have difficulty using public transportation conveniently, and developed countries and least developed countries also need to

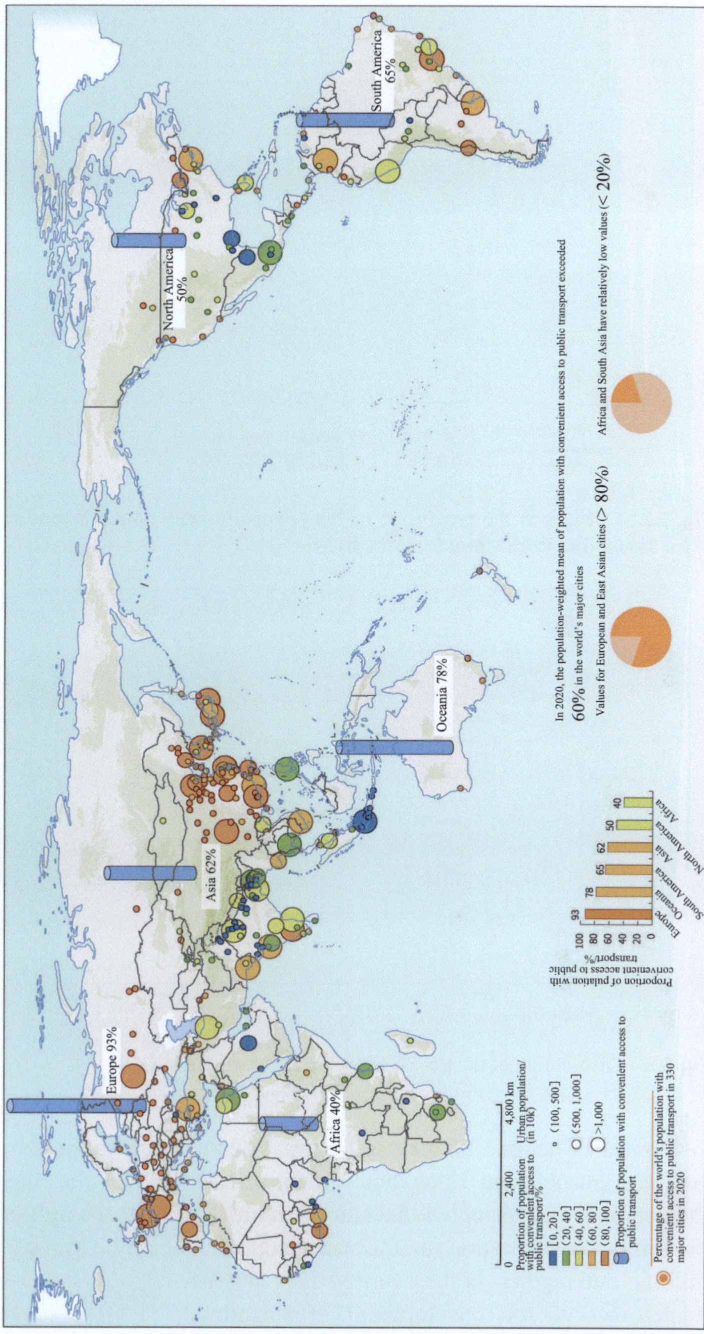

Fig. 5.1 Proportion of population with convenient access to public transportation in 330 major cities around the world in 2020

improve the conditions for about 60 million people to use public transportation.

The proportion of the population in China that can conveniently use public transportation has increased significantly. Among cities with a population of one million, the proportion of people who can conveniently use public transportation has exceeded 50%, mainly concentrated in the Yangtze River Delta and the Pearl River Delta (Fig. 5.6). The top 10 cities in China with the most people who can conveniently use public transportation in 2023 were Shanghai, Beijing,

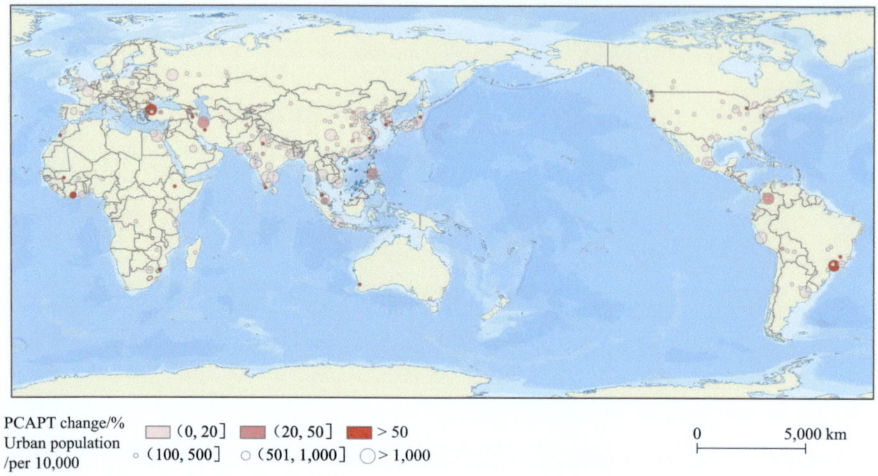

Fig. 5.2 Changes in the proportion of the population with convenient access to public transportation in 330 major cities around the world from 2015 to 2020

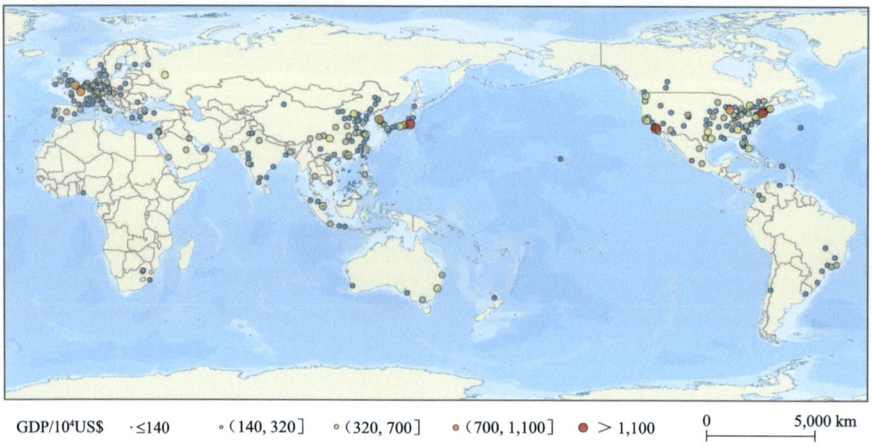

Fig. 5.3 Distribution of the top 300 cities in terms of global GDP

Guangzhou, Chengdu, Shenzhen, Tianjin, Wuhan, Nanjing, Suzhou and Hangzhou. Among them, the population in Shanghai that can conveniently use public transportation in 2023 was about 27.8 million, ranking first in the country. This may be related to economic development, urbanization, and population density.

China's public transportation development level is generally good, but regional development is uneven. Cities with convenient public transportation populations are geographically concentrated in the eastern region. Among them, Shanghai and Beijing have more than 20 million people who can conveniently use public transportation, ranking at the top. Guangzhou, Chengdu, Shenzhen, and Tianjin also have more than 10 million people who can conveniently use public transportation (Fig. 5.7).

Highlights

- This case study produced a dataset on the population distribution with convenient access to public transportation in 330 major cities worldwide in 2015 and 2020 to address the data missing issue for SDG 11.2.1.
- In 2020, the proportion of the population in major cities around the world who can conveniently use public transportation exceeded

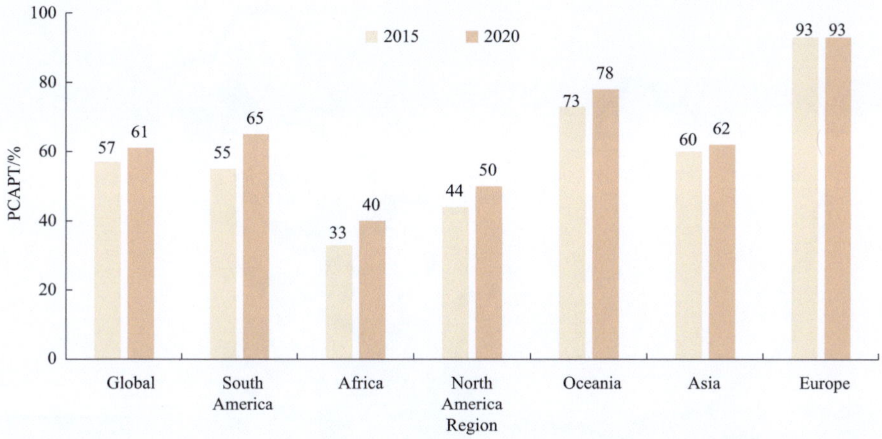

Fig. 5.4 Proportion of the population with convenient access to public transportation in major cities across the world

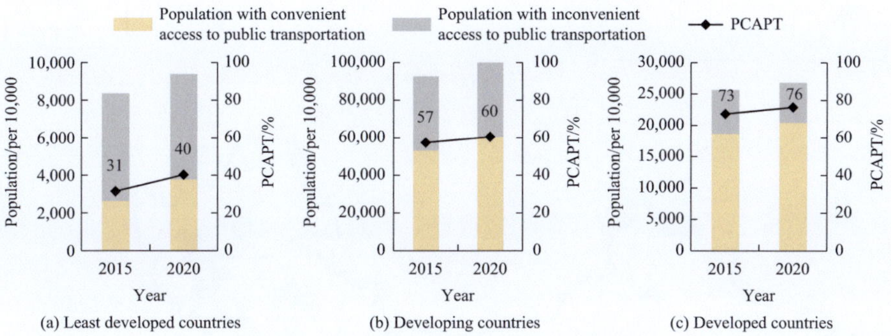

Fig. 5.5 Proportion of the population with convenient access to public transportation and its changes in major cities of different types of countries around the world

60%, an increase of 3.38% compared with 2015; the Gini coefficient was 0.59 (basically the same as in 2015), indicating that the proportion of the population in cities around the world who can conveniently use public transportation is still relatively unbalanced. Through an in-depth analysis of the current state of public transportation accessibility in different regions, this case study provides a valuable reference for urban planners and policymakers in formulating more targeted and effective improvement measures, such as promoting public transportation infrastructure development, enhancing service quality, and expanding coverage.

- **Discussion and Outlook**

Based on global public transportation data, global population grid data, and GHSL-UCDB data, this case study produced a 100 m spatial resolution population distribution dataset of major cities worldwide that can conveniently use public transportation, revealing the spatial patterns and changes of PCAPT indicators at global and Chinese scales, which can provide a reference for local governments to improve public transportation infrastructure. This case study designed a quantitative method to calculate the proportion of the urban population that can conveniently use public transportation systems.

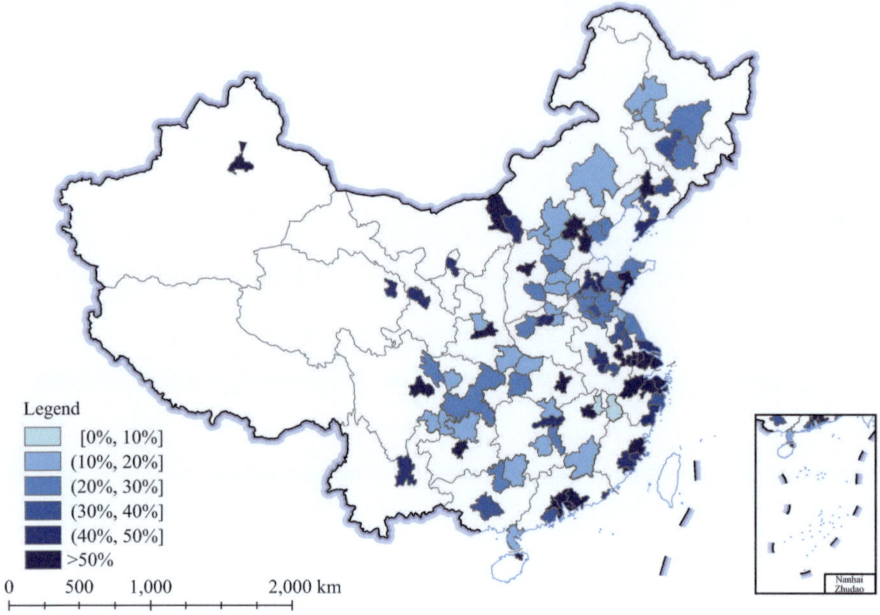

(a) 2017

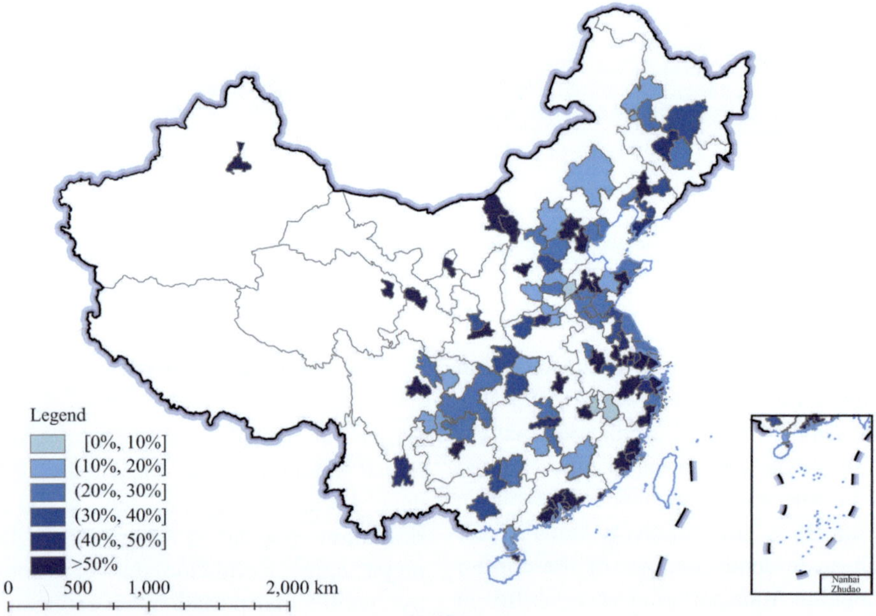

(a) 2023

Fig. 5.6 Proportion of the population in Chinese cities with a population of one million that can conveniently use public transportation in 2017 and 2023

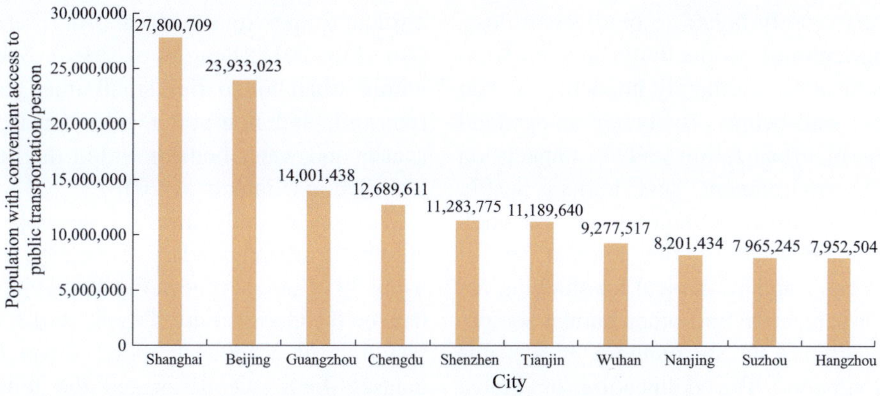

Fig. 5.7 Top ten cities by population with convenient access to public buses in 2023

By integrating multi-source spatiotemporal data and public transportation accessibility analysis, this case study reveals the degree of convenient use of public transportation systems by populations in different cities around the world, taking into account regional and temporal changes and assessing service fairness through the Gini coefficient, addressing the core content of SDG 11.2.1 indicator evaluation. The method is reproducible, and the data are easily accessible, providing a supportive framework for countries worldwide to conduct evaluations of this indicator. The fairness of accessibility has been noted in this case study, but this is mainly based on evaluating accessibility indicators between different regions. In fact, even within the same region, different gender and age groups have varying levels of access to the convenience of public transportation. For example, vulnerable groups such as the elderly and infants find it difficult to use the public transportation system even if they are within the range of convenient public transportation (Ravensbergen et al. 2022). To more scientifically and reasonably evaluate the proportion of the urban population that can conveniently use public transportation, the evaluation results classified by gender and age will be analyzed in the future.

5.3.2 Global Urban Boundary Spatial Distribution Product

Target: SDG 11.3 By 2030, enhance inclusive and sustainable urbanization and capacity for participatory, integrated and sustainable human settlement planning and management in all countries.

- **Background**

Currently, more than 50% of the world's population lives in urban areas, and this proportion is expected to reach 70% by the middle of the twenty-first century (United Nations et al. 2019). One of the direct consequences of urban population growth is the rapid expansion of urban areas, which has led to the loss of large tracts of farmland and forests, creating significant obstacles to achieving the SDGs, especially in developing regions such as Asia and Africa (Mbow et al. 2017; Xi et al. 2016). Furthermore, the rapid intensification of global urbanization has brought about a growing number of issues, including an increase in the number of slum dwellers, overburdened infrastructure such as roads and transportation systems, worsening air pollution, and unplanned urban sprawl. These problems are closely linked to biodiversity loss, reduced agricultural productivity, and environmental degradation, ultimately impacting human health and well-being. To better understand the process of urbanization and its impacts on society, the environment, and human health, clearly defined urban boundaries are essential. Urban boundaries are a fundamental spatial attribute of cities and are critical for studying the effects of urbanization and other human socioeconomic activities on surrounding ecosystems and environments. The delineation of global urban boundaries (GUB) not only helps to define the extent of urban areas but also provides an important basis for land-use planning and resource management during urbanization. Therefore, strengthening research and practice on the delineation of global urban boundaries is of great significance in addressing the challenges posed by urbanization and promoting sustainable urban development.

In most previous studies on global urban boundary extraction, satellite data has been the primary data source. For instance, representative urban extents have been identified and extracted using nighttime light (NTL) data from the Defense Meteorological Satellite Program (DMSP), as well as data from MODIS, Landsat, Sentinel, and SAR (Elvidge et al. 2007; Li et al. 2018; Gong et al. 2013; Esch et al. 2013). However, due to differences in satellite launch times and detection methods, the urban extents derived from these different satellites vary in terms of dates and regions. Additionally, definitions of urban boundaries differ across various application contexts, which limits their utility in global change research. To address these issues, we constructed the first global multi-temporal urban boundary dataset based on long-term high-resolution urban extent maps. Generally, the definition of urban boundaries relies on high-resolution information such as roads, rivers, building blocks, and urban districts (Zhang et al. 2019). However, such information is often difficult to comprehensively obtain on a global scale. Considering the diversity of global urban environments, we adopted a widely used method for defining urban boundaries, which is primarily based on the spatial distribution of artificial impervious surfaces from Landsat data (Hu et al. 2015; Peng et al. 2016). Specifically, within urban boundaries, small urban patches are removed, and non-urban areas (such as green spaces and water bodies) within the boundaries are filled in (Liang et al. 2018).

This case study aims to develop a global automatic urban boundary delineation framework using 30 m annual time-series impervious surface data on the Google Earth Engine (GEE) platform, resulting in a multi-temporal urban boundary dataset. First, we delineated the initial urban boundaries by filling non-urban areas within each city using a combination of KDE and cellular automata-based urban modeling methods. Next, we refined the initial urban boundaries in urban fringe areas using morphological dilation and erosion methods to obtain precise urban boundaries. The extracted GUB aligns closely with NTL data and human-interpreted results, and it effectively captures urban extents when compared to high-resolution Google Earth imagery. The GUB dataset generated in this case study provides a clear physical boundary for urban areas, which will facilitate future research on the impacts of urbanization on food security, biodiversity, climate change, and urban health.

- **Data**

– The 30 m global artificial impervious area (GAIA) annual dynamic dataset from 1985 to 2022 has an average overall accuracy exceeding 90% over multiple years, with particularly notable performance improvements in arid regions. Moreover, the temporal trends of GAIA align well with other datasets at both regional and global scales, making it the only long-term time-series dataset spanning more than 30 years.

- **Method**

In this case study, the KDE method was first used to generate a kernel density (ISA-KD) map from the 1 km impervious surface area (ISA)

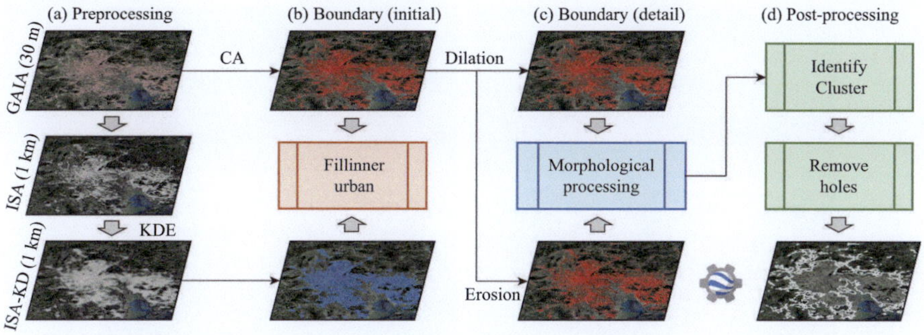

Fig. 5.8 Technical workflow of the global urban boundary dataset

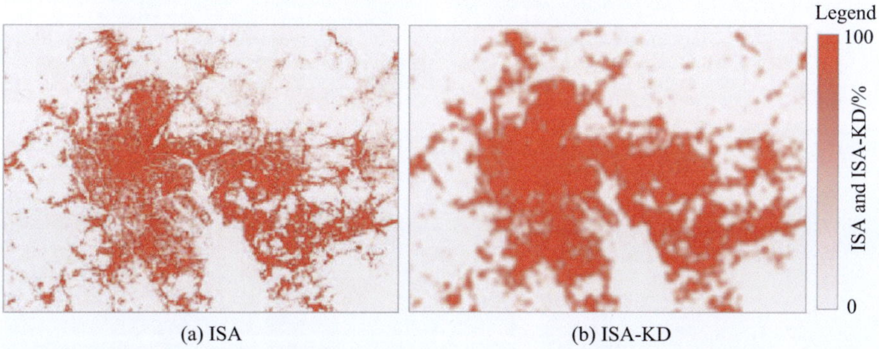

Fig. 5.9 The kernel density map generated from ISA

data aggregated from GAIA. Next, the results of the 30 m resolution urban cellular automata (CA) modeling method were combined with the kernel density map to delineate the initial urban boundary. Subsequently, morphological dilation and erosion methods were applied to refine the urban boundaries in the fringe areas, resulting in the extraction of fine urban boundaries. Finally, some post-processing steps were performed on the generated urban boundaries, such as removing small water bodies and green spaces, to obtain the final urban boundary (Fig. 5.8).

Firstly, the KDE method was used to generate the kernel density map from the 1 km ISA data. Specifically, by aggregating 30 m urban extent data, ISA data was generated (i.e., the percentage of urban area within a 1 km grid). The Asymmetric Gaussian point spread function with a 5 km moving circular window was applied to estimate the kernel density for each ISA raster (Zhao et al. 2020), filling the urban interior areas at a macro scale. The kernel density map provides an approximate range to delineate the urban boundary, as it smooths urban land heterogeneity in central urban areas. Compared to the spatial pattern of ISA, the ISA-KD distribution is more uniform in the central urban areas (Fig. 5.9).

Secondly, the initial urban boundaries were delineated by combining the results of the urban CA modeling method at a 30 m spatial resolution with the kernel density map at a 1 km spatial resolution. Specifically, pixels with ISA-KD values greater than 20% were classified as urban areas (Homer et al. 2015), which filled most of the urban interior regions (Fig. 5.10a). However, urban boundaries generated from coarse-resolution data might miss small urban patches located in or around city centers. To address this, this case study utilized the CA-based method at a 30 m spatial resolution to fill these areas. In the CA model, an 11 × 11 Moore neighborhood was employed, and pixels with a neighborhood density (i.e., the

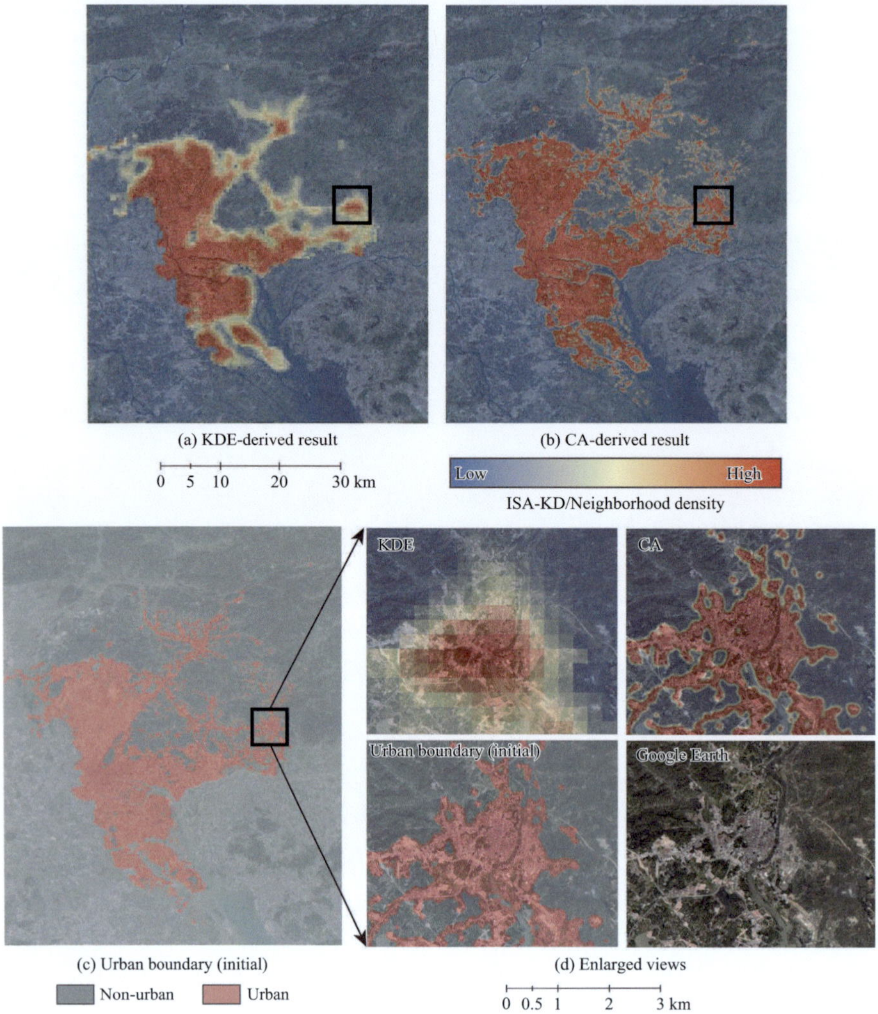

Fig. 5.10 Illustration of delineating an initial urban boundary in Guangzhou, China

percentage of urban pixels within the window) exceeding 20% were classified as urban areas (Kocabas and Dragicevic 2006). The urban CA model was iteratively run three times, resulting in a total expansion distance of approximately 1 km (around 33 pixels) (Fig. 5.10b). By integrating the results of the urban CA model (30 m) and the KDE method (1 km), the initial urban boundaries were established (Fig. 5.10c), which showed strong agreement with the urban extents observed in Google Earth imagery (Fig. 5.10d). Therefore, the combined macro- and micro-scale results effectively filled the most urban interior regions, while the CA modeling method successfully captured the spatial patterns of urban boundaries.

Next, morphological methods were employed to refine the urban boundaries in the surrounding areas. First, a dilation algorithm was applied to merge small urban patches into the main urban extent. Then, an erosion algorithm was used with the same structuring element to remove isolated urban patches. By sequentially applying these two operations, the urban boundaries in the surrounding areas were refined.

Finally, after determining the urban boundaries, post-processing was conducted to refine the generated urban boundaries by removing small "holes", such as water bodies and green spaces, to obtain the final urban boundaries. Specifically, an object-based approach was employed to filter out spatially discontinuous urban clusters (<1 km^2) and eliminate some residual gaps within the urban boundaries, which mainly consisted of green spaces and water bodies that were difficult to completely exclude in previous steps. Through these post-processing steps, the global urban boundary datasets for 2010, 2015, and 2022 were generated and validated by comparative analysis with urban boundary data derived from NTL, manually interpreted results, and official urban area reference data. This process demonstrated the consistency of the GUB with actual urban extents, thereby enhancing the reliability and universality of the results and providing critical data support for sustainable development and management decisions.

- **Results and Analysis**

1. **Urban Boundary Delineation Results**

Our GUB outperforms those from NTL data in terms of the overall pattern of urban boundaries and their spatial details around urban fringe areas (Fig. 5.11). The NTL-derived urban boundary was mapped from annual observations using the DMSP NTL data (Zhou et al. 2018). Available results in these two data sources were compared (i.e., 1995, 2000, 2005, and 2010). Overall, urban dynamics reflected by these two data products agree well (Fig. 5.11a). Compared to the NTL-derived result, GUB can provide more details about urban fringe areas (Fig. 5.11b).

GUB data agree well with the human-interpreted results from Landsat images (Fig. 5.12). Here, GUB in China was compared with human-interpreted results that were also based on Landsat images in three representative years (i.e., 1990, 2000, and 2010) (Wang et al. 2012). We found GUB agreed well with the interpreted boundaries, not only about the overall pattern (Fig. 5.12a) but also the spatial details around urban fringe areas (Fig. 5.12b). However, it should be noted that our approach can: (1) delineate urban boundaries automatically, (2) reduce intensive human labor and subjectivity, and (3) generate consistent results across space and time.

The derived urban sizes from GUB data and reference data in the United States and China are consistent (Fig. 5.13). The urban boundaries in 2010 from the U.S. Census Bureau were digitized with the consideration of population density and land use data within the territory of each census block (https://www.census.gov/). The urban boundaries from interpreted results in China were generated based on Landsat images in 1990, 2000, and 2010 (Wang et al. 2012). In general, urban areas from this case study and reference datasets were distributed around the 1:1 line,

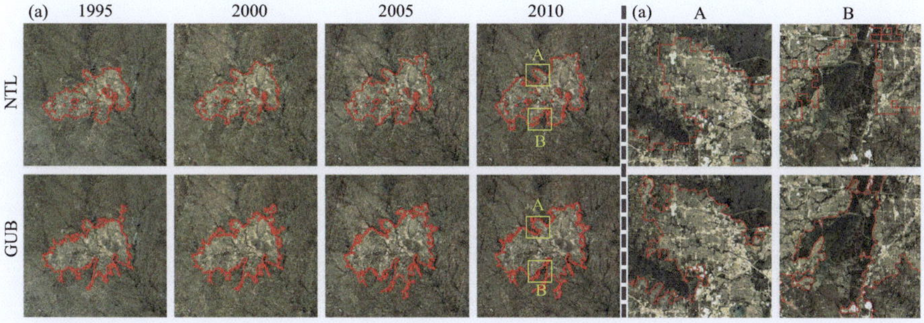

Fig. 5.11 Comparison of the urban boundaries derived from GUB and NTL data in Dallas (United States) at multiple periods (**a**) and their enlarged views at the urban fringe areas in 2010 (**b**). *Note* Spatial extents of each subplot in (a) and (b) are 150 km × 150 km and 15 km × 15 km, respectively. Each image is composited by Landsat data using a natural color (R: Band 3, G: Band 2; B: Band 1)

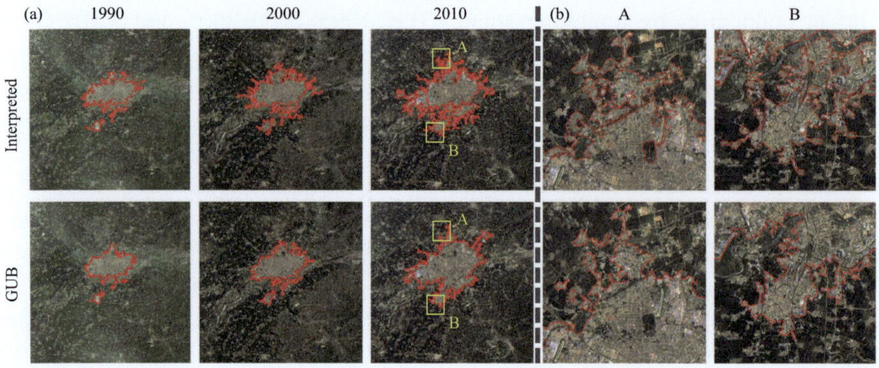

Fig. 5.12 Comparison of the urban boundaries from GUB and interpreted results in Shenyang, China at multiple periods (**a**) and their enlarged views at the urban fringe areas in 2010 (**b**). *Note* Spatial extents of each subplot in (a) and (b) are 60 km × 60 km and 10 km × 10 km, respectively. Each image is composited by Landsat data using a natural color (R: Band 3, G: Band 2; B: Band 1)

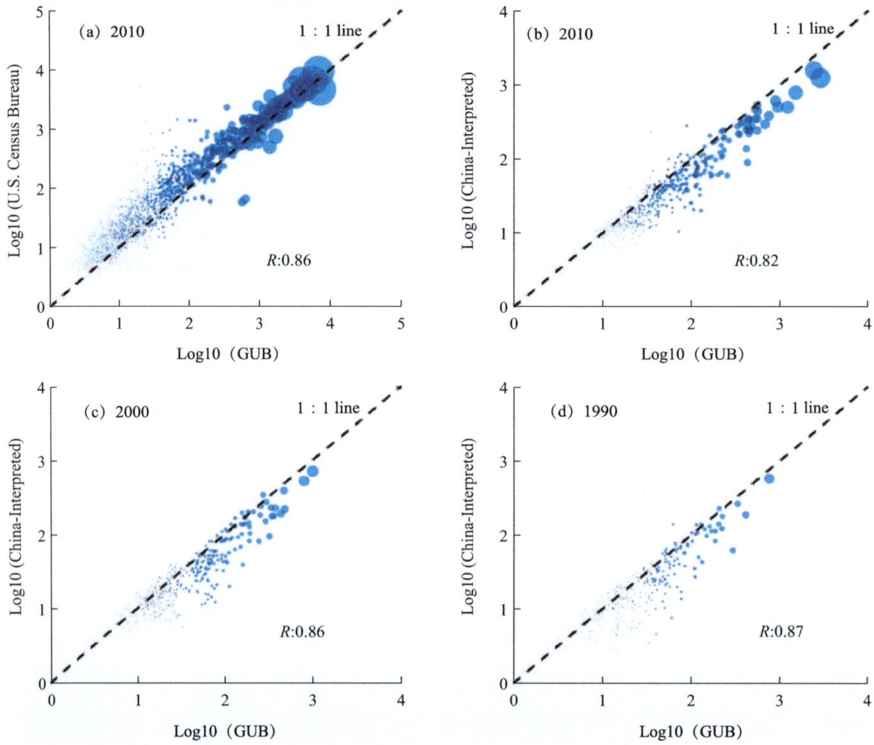

Fig. 5.13 Comparison of urban sizes (km^2) from this case study and the reference datasets. *Note* The relationship of urban size between GUB and the U.S. Census Bureau in 2010 (**a**), and the interpreted results in China in 2010 (**b**), 2000 (**c**), and 1990 (**d**)

with a mean correlation value greater than 0.8. We found urban sizes in our results were slightly smaller than those from the U.S. Census Bureau (Fig. 5.13a), whereas they were slightly higher than those from the interpreted results in China (Fig. 5.13b–d). This is likely attributed to different definitions and approaches used in delineating urban boundaries in these two reference datasets.

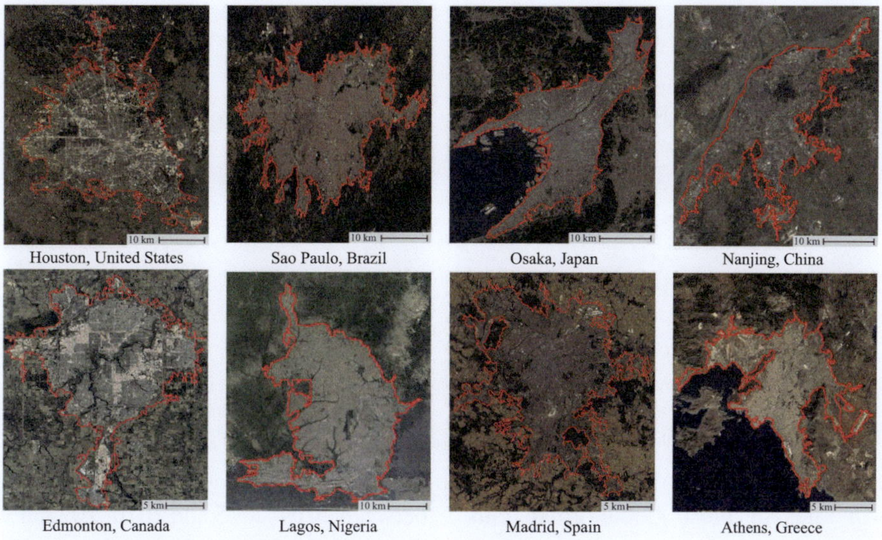

Fig. 5.14 Delineated urban boundaries in eight selected cities in 2022

In our results, the mapping approach was consistent over regions and across years. Besides, the correlation between our and interpreted results in China showed a slightly decreasing trend from 1990 to 2010 (i.e., the correlation coefficient decreased from 0.87 to 0.82), indicating the urban landscape in the urban fringe areas became more complicated due to the rapid urban expansion.

GUB can capture well the spatial extent of urban areas across cities globally (Fig. 5.14). Through overlaying the derived boundaries on high-resolution Google Earth images, we can find that our mapped results match well with the actual urban extent from satellite images. That is, the boundary can well separate urban and surrounding non-urban areas. In addition, these cities are different in terms of the shapes of their derived urban boundaries. For example, cities in North America (e.g., Houston and Edmonton) are more aggregated than cities in Asia (e.g., Nanjing), where there are different branches at urban fringe areas. This suggests that the connections between urban and rural areas in Asia are closer.

In addition, GUB can capture well the dynamics of urban extents (Fig. 5.15). The urban extents in Hefei (China) and Bangkok (Thailand) experienced an evident growth from 2010 to 2022. The delineated urban boundaries include most impervious areas and exclude small urban patches around the urban fringe areas (Fig. 5.15a). This suggests that the developed delineation approach works well for moderate-resolution urban extent maps, which have a relatively high degree of heterogeneity in cities. By overlaying the delineated urban boundaries on satellite images, we found the derived results were reliable and matched well with the actual urban extent (Fig. 5.15b).

2. **Area Change Characteristics of Different Urban Clusters Worldwide**

From a global perspective, urban growth was relatively slow between 2010 and 2015, but urban areas expanded dramatically after 2015, a trend particularly evident in developing countries (Fig. 5.16). By calculating the area growth rates of different urban clusters from 2010 to 2015 and 2010 to 2022, it was found that the eastern coastal region of China consistently showed the fastest urban growth. Additionally, some developing countries, such as Indonesia, India, and parts of Africa, also experienced significant urban area expansion in the later period (Fig. 5.16a, b). Between 2010 and 2015, the

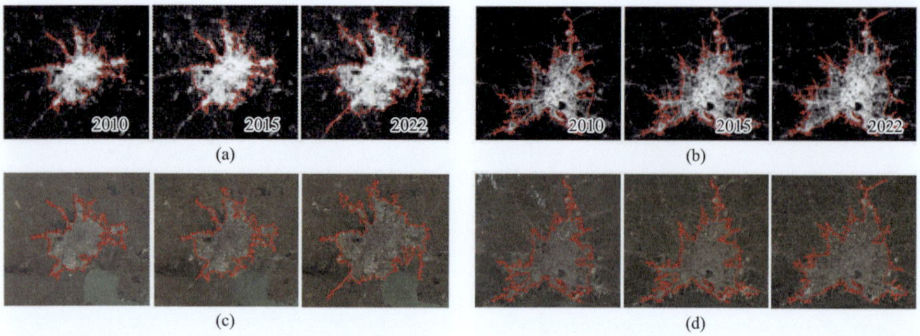

Fig. 5.15 Temporal dynamics of delineated urban boundaries in Hefei (China) on urban extent maps (**a**) and Landsat images (**c**). Similarly, case of Bankok (Thailand) with delineated boundaries and Landsat images are shown in (**b**) and (**d**), respectively. *Note* The spatial extent is 60 km × 60 km. Each Landsat image in (b) was composited using a natural color band combination strategy (i.e., R: Band 3, G: Band 2; B: Band 1)

area growth rates of most global urban clusters exhibited minor differences, with stable or slightly increasing trends, and some even showed contraction, such as in certain areas of Africa and South America. However, urban clusters in eastern China demonstrated a clear trend of substantial growth, especially in coastal provinces like Jiangsu and Zhejiang, where urban growth rates reached as high as 200%. Starting in 2015, urban clusters in regions such as South America, Africa, and Europe expanded rapidly, leading to more pronounced spatial differences in area growth rates of urban clusters from 2010 to 2022.

In addition, this case study conducted a comparative analysis of urban cluster area growth rates across different countries. Compared to Global North Countries, Global South Countries experienced faster urban growth, with more pronounced differences among nations (Fig. 5.16c). This can be attributed to the fact that development in Global North Countries has largely reached saturation, while Global South Countries have experienced rapid development since 2010, particularly after 2015. Countries like China, Thailand, and Indonesia, driven by accelerated economic growth, saw significant urbanization and notable expansion in urban areas. However, disparities in development levels between countries remain substantial, with some countries, such as China, experiencing exceptionally rapid growth. By comparing the area growth rates of urban clusters in the top 10 countries between 2010–2015 and 2010–2022, it was found that countries like the United States and Australia consistently ranked lowest in urban area growth during both periods, while Global South Countries such as China, India, and Brazil led the rankings in the later period. Notably, China's urban growth rate was nearly twice that of the United States and Australia (Fig. 5.16d).

Highlights

- This case study independently produced a multi-temporal vector dataset of global urban cluster extents for the years 2010, 2015, and 2022, providing a fundamental spatial unit for various applications such as urban planning and urban development assessment. This datasat is shareable and widely accessible.
- The GUB dataset demonstrates strong consistency with other products. Comparisons with urban boundaries derived from NTL data show that the GUB-defined urban boundaries effectively capture the complex shapes of urban extents in peripheral areas and align well with manually interpreted results. Additionally, changes in the GUB dataset correspond to the dynamic changes reflected in Landsat time-series data. Fully leveraging the spatiotemporal evolution information of urban extents facilitates the analysis of differences in urban area change rates at various scales and the identification of sensitive regions worldwide,

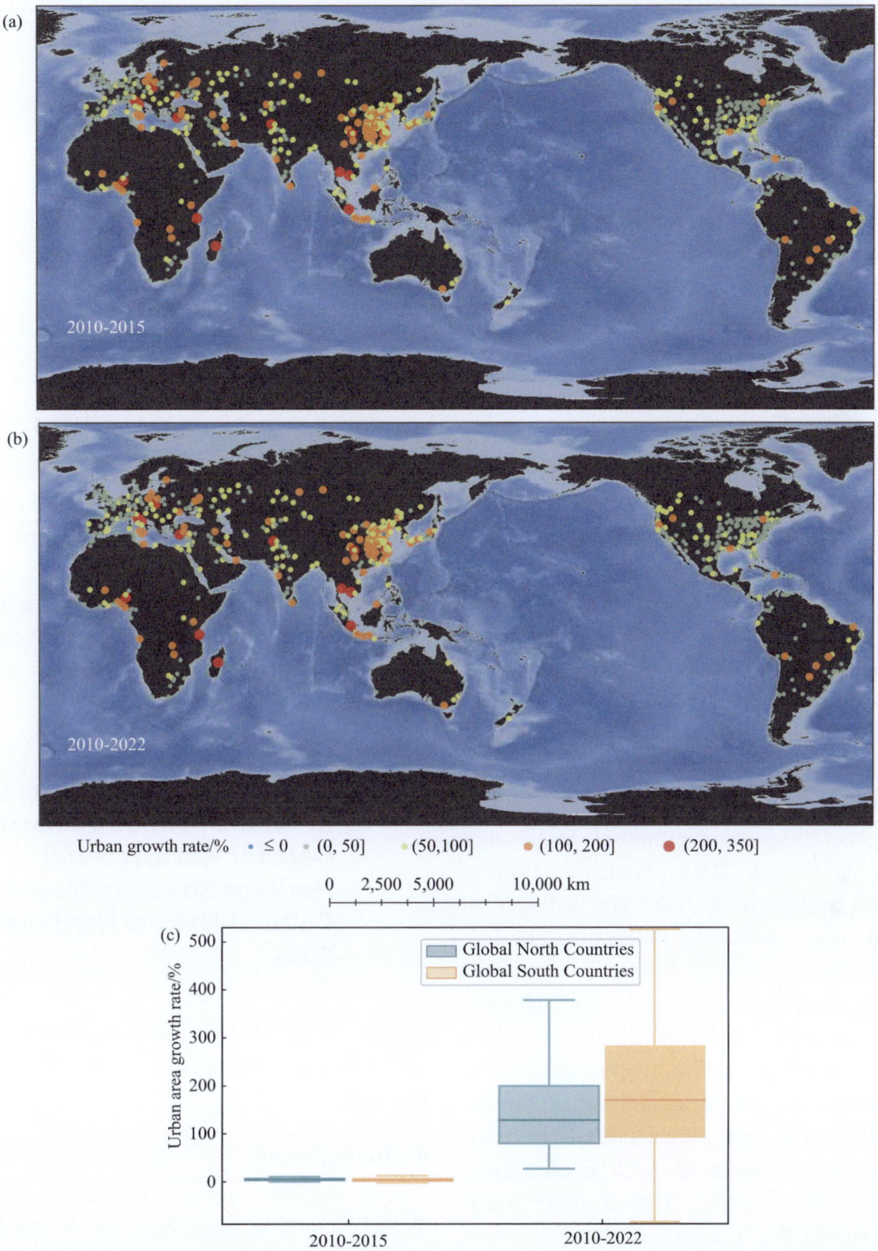

Fig. 5.16 Area growth rates of global urban clusters from 2010 to 2015 (**a**) and from 2010 to 2022 (**b**), area growth rates of urban clusters in Global South Countries and Global North Countries during 2010–2015 and 2010–2022 (**c**), and a comparison of urban area growth rates in the top 10 countries during 2010–2015 and 2010–2022 (**d**)

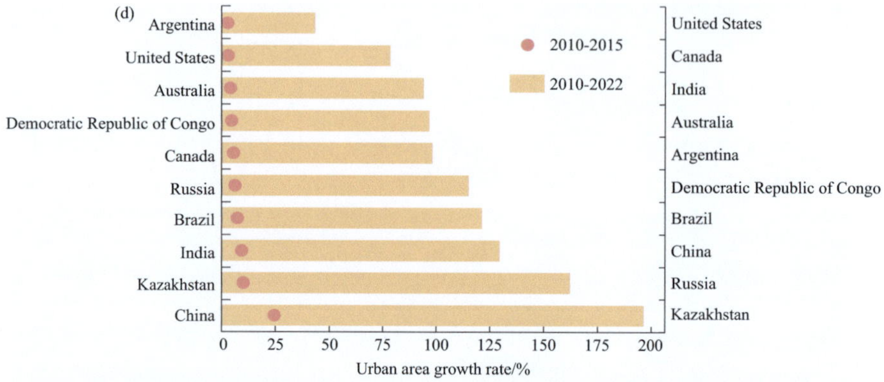

Fig. 5.16 (continued)

providing crucial data support for sustainable development and management decision-making.

- **Discussion and Outlook**

This case study developed a method to delineate urban boundaries from the GAIA dataset without relying on auxiliary datasets or complex models, generating multi-temporal GUB data for 2010, 2015, and 2022. First, we utilized a combined approach of KDE and urban CA-based modeling to define the initial urban boundaries of urban clusters by filling in non-urban areas within the clusters. Second, morphological methods were applied to refine the urban boundaries in peripheral areas. Finally, post-processing was conducted on the generated urban boundaries to obtain the final urban boundaries. The urban boundaries delineated in GUB effectively capture the complex geometries of urban fringe areas and reflect the true extent of urban areas. Moreover, the temporal changes in GUB align well with the dynamic urban extent captured by Landsat. This dataset demonstrates its potential in various urbanization-related studies, such as sustainable planning and ecological conservation. Compared with other products, such as urban extents derived from NTL data, our results excel in detailing urban fringe areas. Given that peripheral urban areas have the greatest impact on biodiversity and agricultural ecosystems due to their vulnerability, these areas are also the most sensitive regions. Considering the dynamic nature of urban development, we will regularly update the dataset to support future research in various urban domains, such as the impacts of urbanization on urban ecosystems, urban landscapes, and urban climates.

5.3.3 Update and Improvement of Global Public Boundary Data for World Natural Heritage Sites and Mixed Cultural-Natural Heritage Sites

Target: SDG 11.4 Strengthen efforts to protect and safeguard the world's cultural and natural heritage.

- **Background**

World Natural Heritage Sites and World Heritage-Mixed Sites are non-renewable resources that integrate biological, ecological, and cultural elements, and possess multiple attributes such as economic, social, and environmental aspects (Wang et al. 2015). SDG 11 aims to create inclusive, safe, resilient, and sustainable cities and human settlements, through seven sub-goals, including SDG 11.4, which calls for "Strengthen efforts to protect and safeguard the world's cultural and natural heritage". The protection of world heritage is a shared global responsibility.

Boundaries are the foundation of natural heritage protection and the fundamental basis for its management. Article 99 of the "Operational Guidelines for the Implementation of the World Heritage Convention" states: "The delineation of boundaries is an essential requirement in the establishment of effective protection of nominated properties. Boundaries should be drawn to incorporate all the attributes that convey the Outstanding Universal Value and to ensure the integrity and/or authenticity of the property." (UNESCO and World Heritage Center 2023) However, due to inconsistencies in the format and precision of boundary data submitted during the early nomination processes of various heritage sites, as well as information asymmetry between the World Heritage Convention (WHC) of the United Nations Educational, Scientific and Cultural Organization (UNESCO) and other heritage monitoring and protection organizations, there are issues of incompleteness, inconsistency, and even inaccuracy in the vector and attribute data from different data sources. The World Database on Protected Areas (WDPA), established in 1981, is the most comprehensive global database on terrestrial and marine protected areas. The vector and attribute data for World Natural Heritage Sites and World Heritage-Mixed Sites in the WDPA require systematic updating and improvement.

The issues in the publicly available data on World Natural Heritage Sites and World Heritage-Mixed Sites pose significant challenges to the protection, management, research, and presentation of heritage values. In response, the International Centre on Space Technologies for Natural and Cultural Heritage under the auspices of UNESCO (HIST) and the International Union for Conservation of Nature (IUCN) have initiated efforts to update and improve the quality of boundary data for World Natural Heritage Sites and World Heritage-Mixed Sites. This work utilizes methods and technologies such as web crawling, natural language processing, and GIS for multi-source data spatial and logical consistency checks. This case study represents a phase of the project outcomes.

- **Data**

– The data utilized in this analysis include fundamental information on 257 World Natural Heritage Sites and World Heritage-Mixed Sites (brief overviews, selection criteria, descriptions of authenticity and integrity, management and conservation requirements), documents related to the nomination and maintenance of these heritage sites (nomination files, evaluation reports from advisory bodies, decision documents, periodic reports, conservation status reports), and non-vector format maps containing spatial information about the heritage sites. All data are current as of April 2023.
– Vector graphics and attribute data for 250 World Natural Heritage Sites and World Heritage- Mixed Sites were extracted from the WDPA, with data current as of April 2023.

- **Method**

The selection, web scraping, and organization of data related to World Natural Heritage Sites and World Heritage-Mixed Sites were conducted as follows. By filtering on the WHC website using the combined criteria of "Natural" and "Mixed", a total of 257 World Natural Heritage Sites and World Heritage-Mixed Sites were identified as of April 2023. Additionally, in the April 2023 edition of the WDPA, filtering the DESIG_ENG (English Designation Name) field for "World Heritage Site (Natural or Mixed)" yielded 250 such sites. Subsequently, web scraping techniques were employed to retrieve documents and map materials related to the nomination and maintenance of these heritage sites. Natural language processing methods were utilized to extract relevant attribute information from the WHC website and Wikipedia pages. The attribute information, vector data, nomination and maintenance documents, and non-vector format maps containing spatial information of the heritage sites were then organized and summarized according to the specific circumstances of each site.

Verification of World Natural Heritage Sites and World Heritage-Mixed Sites in China on the WHC Website was conducted as follows. First, map verification was performed by referring to Articles 6, 8, and 9 of the "Specification for Inspection, Acceptance and Quality Assessment of Digital Surveying and Mapping Products" (GBT 18,316—2001). The map review process included the following steps: (1) checking map format; (2) assessing map clarity; (3) verifying the completeness of map annotations; and (4) confirming the cartographic date. Spatial information was restored using coordinate matching and geometric correction techniques. Second, archival information was reviewed by employing natural language processing methods to extract spatial information related to the heritage site's protected area from the nomination and maintenance records.

Verification of the completeness and accuracy of vector and attribute data for Chinese heritage sites in the WDPA involved the following steps. First, the WDPA vector data for the heritage sites were overlaid with corresponding map sheets to qualitatively describe the degree of overlap, thereby assessing the accuracy of the vector boundary spatial information. Second, the attribute table of the heritage site vector data was checked for completeness and correctness of the attribute items and contents. Specifically, the following were verified: (1) whether any attribute items were missing; and (2) whether the attribute contents were correct. Particular attention was given to checking whether the recorded nominated area in the attribute items matched the information on the WHC website and whether the nominated area was consistent with the GIS area of the graphic. The area difference rate between the nominated area and the GIS area of the graphic was calculated using the formula:

$$S_c = \frac{S_g - S_r}{S_r} \times 100\%$$

where, S_c represents the area difference rate, S_g denotes the GIS area of the graphic, and S_r is the nominated area.

- **Results and Analysis**

The quality and precision of the boundary data for 18 World Natural Heritage Sites and World Heritage-Mixed Sites in China were examined, with the results presented in Table 5.2: (1) Three heritage sites had accurate public archival information that was consistent with the vector data, namely Tianshan in Xinjiang, Shennongjia in Hubei, and Hoh Xil in Qinghai; (2) Fifteen heritage sites had issues with the content of the map files, spatial reference information, or map annotation information on the WHC website, requiring updates. Among these, four heritage sites had correct boundary vectors and attribute data, while 11 sites could not accurately determine the correctness of the boundary vector data due to issues such as missing map files, low resolution, lack of spatial reference information, or significant differences with vector graphics. This verification, using China as an example, revealed numerous issues with the content of heritage site map files and spatial reference information on the WHC website. These issues may also exist for many more heritage sites globally, affecting the verification and updating of their vector data. Since the official website archives are the authoritative source and reference for heritage site information, it is imperative to check and update the information on the WHC website while updating the vector data.

Based on the extraction of 250 heritage site data from the WDPA (as of April 2023), the area difference rate between the nominated area in the official archival materials and the vector boundary GIS area was calculated. The absolute value of the area difference rate was within 10% for 80.5% of the sites, while 50 sites had an absolute difference rate greater than 10%, with the maximum difference rate reaching 513.44%, as illustrated in Figs. 5.17 and 5.18. Heritage sites with larger absolute difference rates are predominantly located in East and Southeast Asia. In China, five heritage sites have an absolute area

Table 5.2 Results of data quality and accuracy examination of the boundaries of the 18 World Natural Heritage Sites and World Heritage-Mixed Sites in the China region

Inspection results		Number of heritage sites/item	Heritage site list
Archive information	Vector and attribute data		
Correct	Good consistency and completeness	3	Tianshan in Xinjiang, Shennongjia in Hubei, Hoh Xil in Qinghai
Partial updates required	Good consistency and completeness	4	Jiuzhaigou Valley Scenic and Historic Interest Area, Sichuan Giant Panda Sanctuaries, Migratory Bird Sanctuaries along the Coast of Yellow Sea-Bo hai Gulf of China (Phase I)
Updates required	Data is complete; consistency requires further verification	11	Mount Huangshan, Sanqingshan Mount Taishan, Huanglong Scenic and Historic Interest Area, Mount Emei-Leshan Giant Buddha, Mount Wuyi, Three Parallel Rivers of Yunnan Protected Areas, China Danxia, South China Karst, Chengjiang Fossil Site, Fanjingshan

difference rate greater than 10%, namely the South China Karst, the Three Parallel Rivers of Yunnan Protected Areas, Hoh Xil in Qinghai, Huanglong Scenic and Historic Interest Area, and Mount Emei-Leshan Giant Buddha. These discrepancies warrant significant attention. Accurate GIS data form the foundation for the protection, management, and planning of heritage sites. The WDPA, currently the most comprehensive global database for terrestrial and marine protected areas, may lead to the misallocation of resources (such as funding and manpower) if there are discrepancies between the vector and attribute data of World Natural Heritage Sites and World Heritage-Mixed Sites and the nominated data. This can affect public and international perceptions and awareness of heritage sites, hinder the smooth progress of international cooperation, and even lead to legal issues, such as violations of international conventions or unclear responsibilities in heritage site management. Therefore, there is an urgent need to update and improve the database system.

Currently, this case study has supplemented the missing vector and attribute data for seven heritage sites in the WDPA. Additionally, the boundary data for 18 heritage sites in China, as well as the "Ahwar of Southern Iraq: Refuge of Biodiversity and the Relict Landscape of the Mesopotamian Cities" and the Wadi Rum Protected Area, have been revised and updated as necessary. The latest version of the vector data includes 257 World Natural Heritage Sites and World Heritage-Mixed Sites (Fig. 5.19), making it the most complete and accurate dataset of boundaries for World Natural Heritage Sites and World Heritage-Mixed Sites to date.

Highlights

- In the realm of Big Earth Data, this case study introduces a boundary information extraction method based on natural language processing. It conducts a sampling verification of the boundary data for World Natural Heritage Sites and World Heritage-Mixed Sites, identifying both common and unique issues in the boundary data. The global boundary data for World Natural Heritage Sites and World Heritage-Mixed Sites have been updated (2023 edition), and further recommendations for data improvement have been proposed. This work provides valuable reference and guidance for UNESCO in the enhancement and updating of heritage data.
- A sampling inspection of the quality and precision of boundary data for World Natural Heritage Sites and World Heritage-Mixed Sites reveals that some sites have issues with the

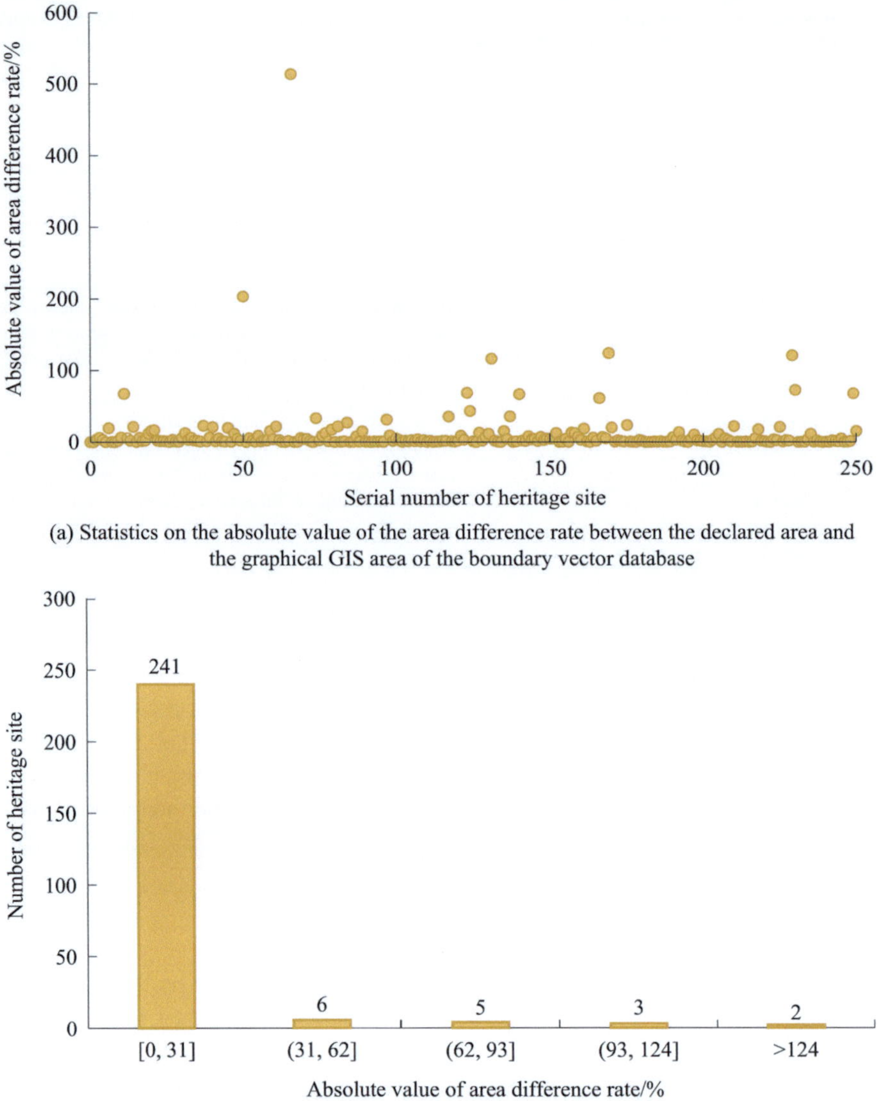

(a) Statistics on the absolute value of the area difference rate between the declared area and the graphical GIS area of the boundary vector database

(b) Distribution of absolute value of area difference rate between declared area and the graphical GIS area of the boundary vector database

Fig. 5.17 Absolute value of area difference rate between declared area and GIS area of boundary vector database graphic

content of their publicly available map files and spatial reference information, which affects the verification and updating of their vector data. Additionally, there are discrepancies between the nominated areas of some heritage sites and their GIS graphic areas in the WDPA. Future efforts will involve collaborating with IUCN to further improve and update the data.

- **Discussion and Outlook**

The boundary information extraction method based on natural language processing, as applied in this case study, was used to extract and organize boundary information for World Natural Heritage Sites and World Heritage-Mixed Sites from the WHC website and related

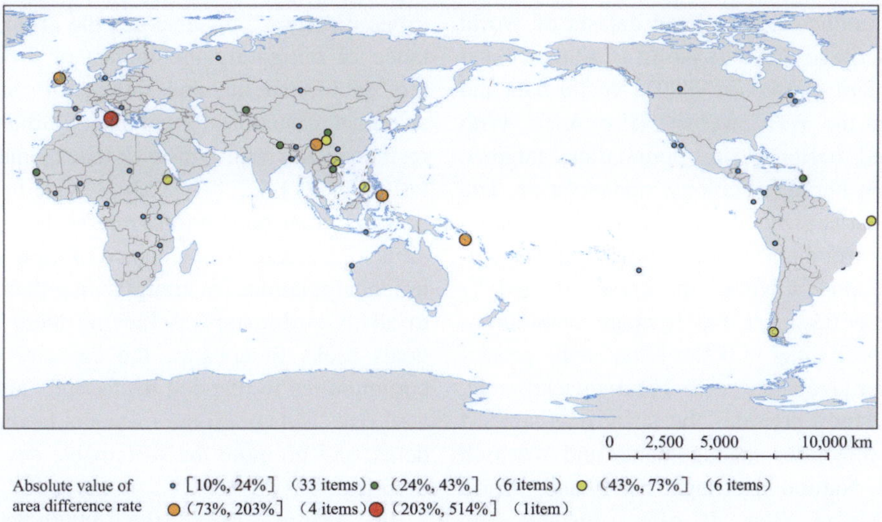

Fig. 5.18 Distribution of the 50 World Natural Heritage Sites and World Heritage-Mixed Sites with an absolute rate of difference of more than 10% in area

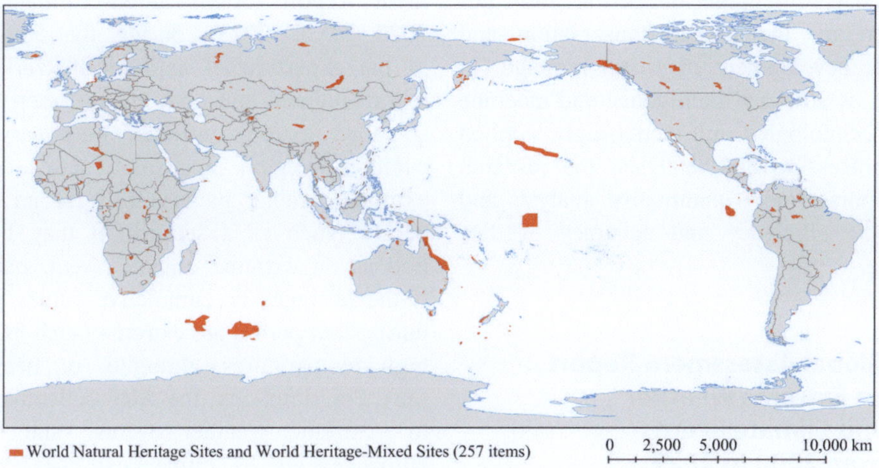

Fig. 5.19 World Natural Heritage Sites and World Heritage-Mixed Sites boundary data (Updated 2023)

English-language websites. Our analysis of the content of map files and spatial reference information related to heritage site boundaries on the WHC website, it was found that out of the 257 World Natural Heritage Sites and World Heritage-Mixed Sites globally, 129 (50%) have complete map elements, 126 have incomplete map elements, and 2 sites lack boundary maps. Therefore, there is an urgent need to improve and update the map and boundary information on the WHC website. Currently, under the collaboration of HIST and IUCN, the heritage site boundary vector data based on the WDPA has been updated. In the future, as the information on the WHC website continues to improve, the quality of the boundary vector database is expected to be further enhanced accordingly.

This case study utilized the boundary information extraction method based on natural language processing to assess the completeness and consistency of boundary information for World Natural Heritage Sites and World Heritage-Mixed Sites. It has identified both common and unique issues in the boundary data and produced an updated dataset of World Natural Heritage Site and World Heritage-Mixed Site boundaries based on WDPA vector data and maps from the WHC website (Fig. 5.19). With the revised heritage site boundaries, targeted research on heritage ecology, conservation, and management is planned, particularly focusing on human–environment relationships and sustainable development within the areas affected by human activities near the heritage boundaries. This approach aims to better balance the protection of heritage sites with local development. This case study provides the latest data support for the protection, management, and research of World Natural Heritage Sites and World Heritage-Mixed Sites. It offers insights and references for UNESCO to improve and update heritage data, thereby aligning with UNESCO's and the SDGs' heritage protection strategies and supporting the global conservation and sustainable development of world heritage. In the future, as artificial intelligence and machine learning technologies are increasingly applied in data processing and analysis, the research will incorporate more quantitative analysis and enhance the efficiency and accuracy of data processing.

5.3.4 Global Assessment Report on Extreme Weather and Climate Events from 2000 to 2022

Target: SDG 11.5 By 2030, significantly reduce the number of deaths and the number of people affected and substantially decrease the direct economic losses relative to global gross domestic product caused by disasters, including water-related disasters, with a focus on protecting the poor and people in vulnerable situations.

- **Background**

SDG 11 focuses on sustainable cities and communities. Among its objectives, SDG 11.5.1, "By 2030, reduce the number of deaths caused by urban disasters", underscores the critical importance of mitigating the impact of urban disasters on human life and safety. This goal aims to reduce casualties by enhancing urban disaster resilience and improving risk management. Similarly, SDG 11.5.2, "By 2030, reduce the proportion of urban populations injured by disasters", aims to decrease the incidence of injuries among urban populations by strengthening their capacity to adapt to disasters. Achieving these two sub-goals seeks to enhance the capacity of urban communities to respond to disasters and associated risks, safeguard the lives and health of residents, and promote the sustainable development of cities.

In March 2023, the Intergovernmental Panel on Climate Change (IPCC) finalized the Synthesis Report of the IPCC Sixth Assessment Report (AR6), titled Climate Change 2023 (IPCC 2023), during the 58th session of the expert panel held in Switzerland. The report pointed out that the characteristics of so-called extreme weather may vary significantly by region in absolute terms. When an extreme weather pattern lasts for an extended period (such as a season), it may be classified as an extreme climate event, particularly if the averages or cumulative values produced during that period are extreme (such as seasonal high temperatures, droughts or heavy rainfall). For simplicity, the AR6 collectively refers to "extreme weather events" and "extreme climate events" as "climate extremes".[1] Climate extremes are one of the seven major scientific challenges identified by the World Climate Research Programme (WCRP) and a key focus of the assessment reports of the IPCC. They are usually characterized by suddenness, high destructiveness, wide-ranging impacts and prolonged duration, bringing significant difficulties and challenges to disaster management efforts worldwide.

[1] Reisinger A, Cammarano D, Fischlin A, et al. Synthesis Report of the IPCC Sixth Assessment Report (AR6). https://www.ipcc.ch/report/ar6/syr/downloads/report/IPCC_AR6_SYR_Annex-I.pdf [2024-10-10].

On May 18–19, 2023, the high-level meeting for the Midterm Review of the Sendai Framework for Disaster Risk Reduction 2015–2030 (Sendai Framework for short) was held at the United Nations Headquarters in New York. During the meeting, the Report of the Midterm Review of the Implementation of the Sendai Framework for Disaster Risk Reduction 2015–2030 (UNDRR 2023) was released and pointed out that the Sendai Framework has driven a global shift since 2015 from disaster management to disaster risk management in decision-making, investment, and action, but the progress data on disaster impact reduction indicate that some countries have not achieved the expected outcomes of significantly reducing disaster losses by 2030 as planned. Moreover, disparities between nations and regions have become more serious.[2] Conducting forward-looking and corrective assessments of climate extremes can provide critical support for reducing vulnerability, risk and inequality.

- **Data**

- Data on global natural disasters (including 5 major categories of droughts, floods, extreme temperatures, storms and other weather disasters) from 2000 to 2022, obtained from the Emergency Events Database (EM-DAT).
- Data on global natural disasters (including 5 major categories of floods, wildfires, droughts, storms and other meteorological disasters) from 2000 to 2022, obtained from the Global Disaster Data Platform (GDDAT).

- **Method**

This case study aligned with the AR6, which collectively referred to "extreme weather events" and "extreme climate events" as "climate extremes", and adopted the technical approach from two-dimensional to multi-dimensional analysis, from extreme value thresholds to process-based events (Fig. 5.20), from static monitoring to spatiotemporal dynamic identification, and from manual identification to objective identification (Fig. 5.21). Based on the criteria for determining climate extremes, multi-source data integration and screening were conducted for global disaster events, with comprehensive consideration of the intensity of hazards and the severity of resulting impacts. To determine a climate extreme, at least one of the following criteria must be met: (1) The intensity of the event ranks within the top 1% of similar disasters worldwide[3]; (2) The number of deaths caused by the event exceeds 500; (3) The number of people affected by the event exceeds 10 million; (4) The direct economic losses caused by the event exceed USD1 billion.

The case study identified 652 climate extremes that met the criteria from 2000 to 2022, including 5 types of disasters, and generated a global climate extreme event dataset from 2000 to 2022. An in-depth analysis of global climate extremes from 2000 to 2022 was conducted, focusing on the spatial and temporal distribution and characteristics of change. Additionally, comparisons were made regarding the frequency of hazards, the number of people affected, the number of deaths, and the direct economic losses of climate extremes before and after the implementation of the Sendai Framework.

- **Results and Analysis**

1. **Global Extreme Weather and Climate Events from 2000 to 2022 Were Dominated by Extreme Storms, Extreme Floods and Extreme Droughts**

[2] UNDRR. Sendai Framework. https://sendaiframework-mtr.undrr.org/publication/report-midterm-review-implementation-sendai-framework-disaster-risk-reduction-2015-2030 [2021-10-10].

[3] According to EM-DAT data analysis, the disaster intensity thresholds for the top 1% of global weather and climate events are: extreme high temperature (52°C), extreme low temperature (−51°C), extreme floods (affected area exceeding 585,146 km^2), extreme droughts (affected area exceeding 32,650 km^2), and extreme storms (maximum wind speed exceeding 320 km/h).

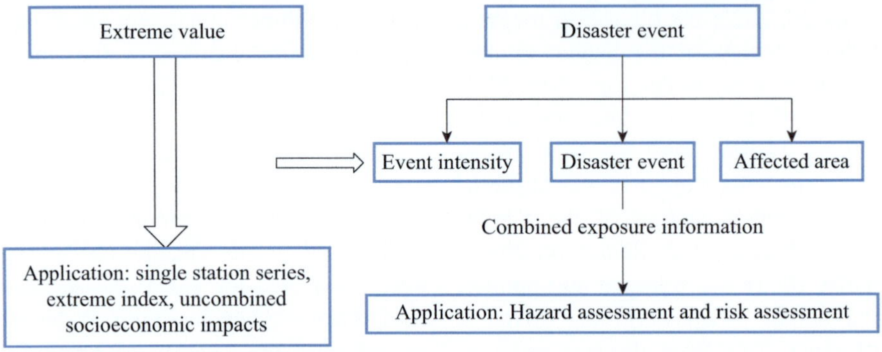

Fig. 5.20 Technical approach for identifying extreme weather and climate events: from extreme value thresholds to process-based events

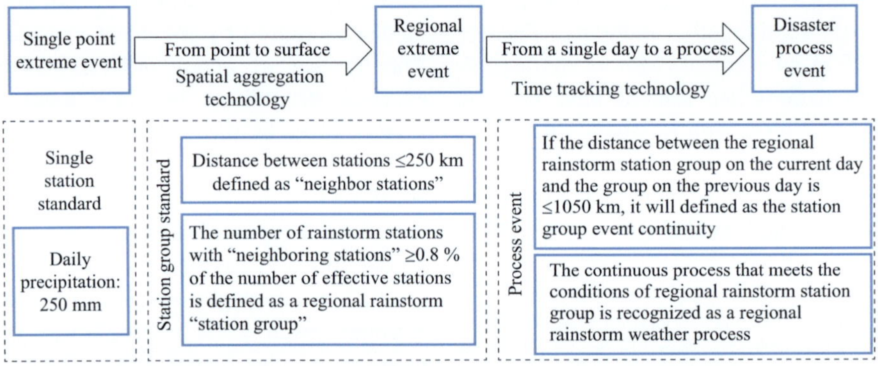

Fig. 5.21 Technical approach for identifying extreme weather and climate events: from manual identification to objective identification

From 2000 to 2022, there were 7,093 weather and climate events in the world, and 652 extreme climate events that met the standards, including 5 types of disasters, namely extreme floods, extreme droughts, extreme high temperatures, extreme low temperatures, and extreme storms (Fig. 5.22). Among them, extreme storms (327 events), extreme floods (209 events) and extreme droughts (66 events) accounted for 50%, 32% and 10% of the total number of global climate extremes, which were respectively the main types of climate extremes that affected the world. From the distribution of various types of climate extremes in various continents around the world from 2000 to 2022, the frequency of climate extremes in the Americas and Asia was relatively high, 255 times and 238 times, respectively, accounting for 39.1% and 36.5% of the total number of global climate extremes. The frequency of extreme climate events in Europe was second only to the Americas and Asia, but the absolute number was only about 40% of that in the Americas and Asia, with a total of 98 events, accounting for 15% of the total number of extreme climate events worldwide. The frequency of extreme climate events in Africa and Oceania was similar, with 31 and 30 events respectively, accounting for 4.8% and 4.6% of the total number of extreme climate events worldwide.

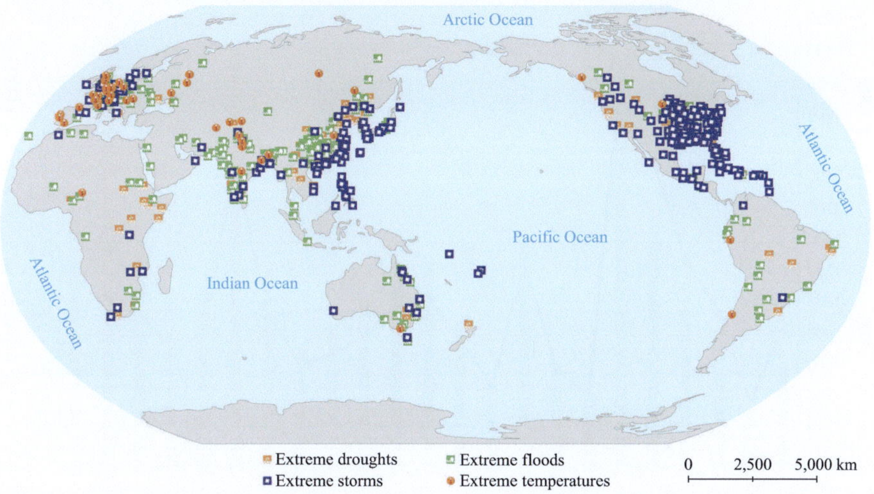

Fig. 5.22 Global spatial distribution of 652 extreme weather and climate events from 2000 to 2022

2. **There Has Been a Notable Reduction in the Average Annual Number of Deaths and Number of Deaths Per 100,000 People Attributed to Extreme Weather and Climate Events Globally Since the Sendai Framework Began To Be Implemented**

After the implementation of the Sendai Framework (2016–2022), the average annual number of deaths due to climate extremes worldwide decreased by 45.4% compared with before its implementation (2000–2015), and the average annual number of deaths per 100,000 people was significantly lower than before its implementation (Fig. 5.23). Specifically, from 2000 to 2015, global climate extremes caused a total of 410,000 deaths, an average of 26,000 deaths per year, and an average of 0.38 deaths per 100,000 people per year; from 2016 to 2022, global climate extremes caused a total of 98,000 deaths, an average of

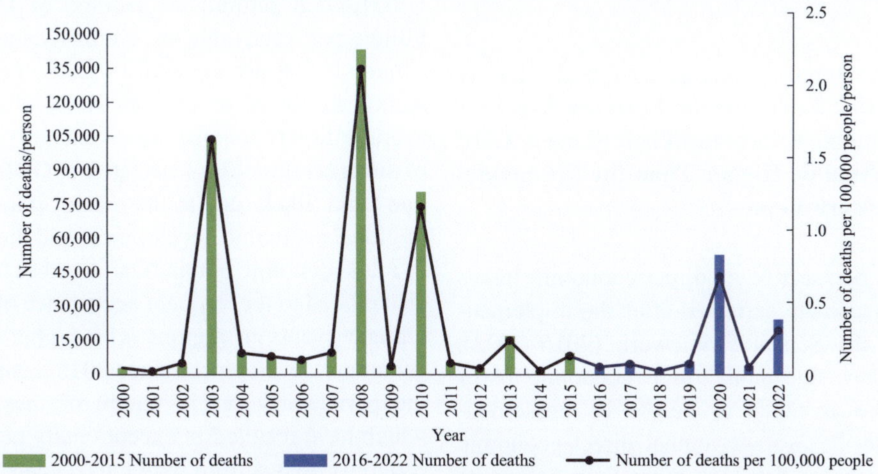

Fig. 5.23 Comparison of global deaths from extreme climate events before and after the implementation of the Sendai Framework. *Note* The number of deaths per 100,000 people refers to the proportion of deaths in that year to the total global population in that year. The global total population data comes from the WB (https://data.worldbank.org/)

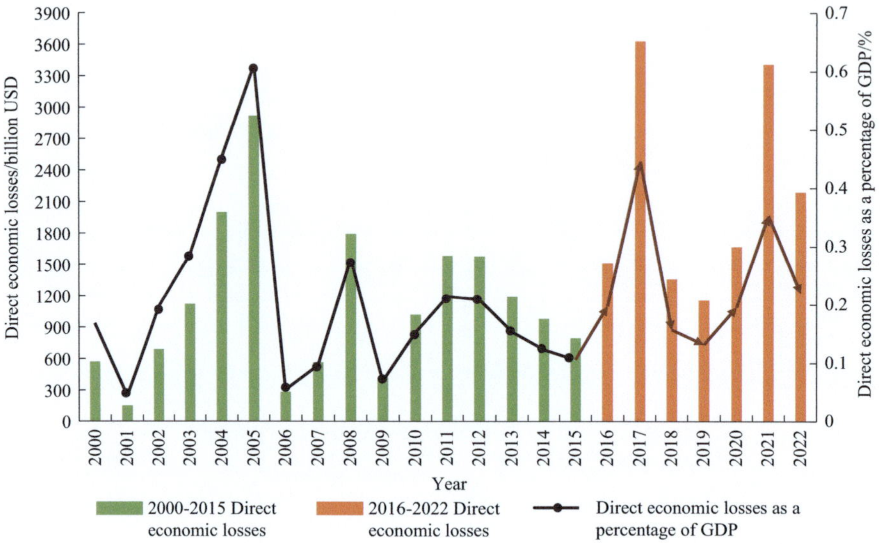

Fig. 5.24 Comparison of direct economic losses from global climate extremes before and after the implementation of the Sendai Framework. *Note* The direct economic loss as a percentage of GDP refers to the total direct economic loss of the year as a percentage of the global GDP of the year. The global GDP (current USD) data comes from the WB (https://data.worldbank.org/), and the 2022 data comes from the IMF (https://www.imf.org/en/Home)

14,000 deaths per year, and an average of 0.18 deaths per 100,000 people per year. It can be seen that after the implementation of the Sendai Framework, the global average annual number of deaths and the number of deaths per 100,000 people have both decreased significantly.

3. **Both Global Average Annual Direct Economic Losses from Extreme Weather and Climate Events and Their Share of GDP Were Notably Higher Than the Pre-Sendai Framework Levels**

From the perspective of direct economic losses caused by climate extremes after the implementation of the Sendai Framework (2016–2022), the global average annual direct economic losses were twice as much as before its implementation, and the average annual direct economic losses accounted for 21% more than before its implementation[4] (Fig. 5.24). Specifically, from 2000 to 2015, global climate extremes caused a total direct economic loss of USD1,764.7 billion, an average of USD110.3 billion per year, and an average annual direct economic loss accounted for 0.2% of GDP; from 2016 to 2022, global climate extremes caused a total direct economic loss of USD1,492.8 billion, an average of USD213.3 billion per year, and an average annual direct economic loss accounted for 0.24% of GDP. The significant increase in both the global annual average direct economic losses and the proportion of direct economic losses relative to GDP is, on the one hand, likely due to the overall improvement in global economic development and the increase in total economic output. On the other hand, it is also related to the frequent occurrence of extreme climate events in regions with higher levels of economic development since 2016, coupled with the particularly severe nature of these events, which have resulted in exceptionally large direct economic losses from individual events.

[4] The average annual direct economic loss as a percentage of GDP is equal to the average of the total direct economic losses in each year as a percentage of the global GDP in that year.

Highlights

- Using the data from EM-DAT and GDDAT, this case study compiled and integrated a global climate extreme dataset from 2000 to 2022, which addressed gaps in the data required for SDG 11.5.1 and SDG 11.5.2, available for sharing.
- From 2000 to 2022, Asia experienced the highest number of fatalities caused by climate extremes, totaling 223,000 deaths, with an average annual death toll of 9,685. In contrast, the Americas incurred the highest direct economic losses, amounting to USD1,932.9 billion, with significant annual variation, reaching USD333.3 billion in 2017, USD252.6 billion in 2005, and USD246.1 billion in 2021. These figures highlight the significant disparities in vulnerability, risk levels and emergency response capacities across global regions when facing threats from climate extremes.
- After the implementation of the Sendai Framework (2016-2022), the frequency of global climate extremes increased, the average annual number of affected individuals rose by 70.3% compared to pre-Sendai levels, and the average annual direct economic losses doubled, with the proportion of direct economic losses relative to GDP increasing by 21%. However, the number of deaths decreased by 46.2% and the average annual death rate per 100,000 people was significantly lower than pre-Sendai levels. These outcomes were of great significance to the realization of the first specific target of the Sendai Framework.[5]

- **Discussion and Outlook**

This case study found that from 2000 to 2022, the direct economic losses caused by climate extremes showed a fluctuating and increasing trend across Asia, the Americas, Europe, Africa and Oceania. In terms of total economic losses, Asia and the Americas were the most severely affected regions during this period. In particular, the economic losses in the Americas exhibited more pronounced extremity, while Asia experienced a higher annual frequency of climate extreme events and more sustained economic losses. In Europe, the impact varied significantly over the years. Although Africa and Oceania accounted for a smaller share of economic losses, Africa ranked among the highest globally in disaster-related fatalities, and Oceania in disaster frequency. These findings illustrate the diverse impacts of climate extremes across regions with varying levels of development and underscore the necessity and urgency of a global response to such disasters.

As global warming intensifies, the risks, foreseeable adverse impacts and associated losses and damages caused by climate change are escalating. Climate and non-climate-related risks are increasingly interacting, resulting in more complex and challenging compound and cascading risks. To mitigate the growing threats posed by more frequent and severe climate extremes under climate change, it is crucial to enhance research on the patterns and mechanisms of climate disasters, improve forecasting and early warning capabilities for frequent climate disasters, and strengthen the monitoring and risk assessment of climate disasters. Additionally, we should pay attention to the heightened risks in agriculture and water resources, escalating ecological security risks, and increased health and safety concerns related to climate extremes in the process of economic construction and regional development.

5.3.5 Global Change in Urban Greenness and Beneficiary Population in Large Cities

Target: SDG 11.7 By 2030, provide universal access to safe, inclusive and accessible, green and public spaces, in particular for women and children, older persons and persons with disabilities.

[5] BY 2030, the global disaster mortality rate will be significantly reduced, and the average global disaster mortality rate per 100,000 people between 2020 and 2030 will be lower than the level between 2005 and 2015.

- **Background**

According to the latest report released by UN-Habitat, the global urbanization rate will increase from 56% in 2021 to 68% by 2050 (UN-Habitat 2022). The sustainable development of cities is directly related to the future of humankind, but differences in socioeconomic development levels have led to an extremely uneven global urbanization process (Sun et al. 2020). In developed countries, municipal green infrastructure is relatively complete, and urban residents can enjoy good green space as public services. Relatively speaking, in many developing countries and underdeveloped countries, cities not only lack basic infrastructure (such as sanitary drinking water, reliable energy supply, health and education resources) but also lack public green space. On a global scale, the continuous dynamic monitoring of urban green spaces in time and space is critical to the achievement of SDG 11.

In 2004, the State Forestry Project Administration launched the "National Forest City". As of January 2024, China's national forest cities had increased to 219, advancing the country's goal of establishing an "ecological civilization" in urban areas. However, quantitative comparative research has been insufficient for identifying the most important aspects of urban greenness in the construction of an urban ecological civilization.

- **Data**

- 2022 MODIS MCD12Q1 land-cover type dataset [International Geosphere-Biosphere Programme (IGBP) classification] at a spatial resolution of 500 m.
- Enhanced vegetation index (EVI) data from the MODIS dataset (MOD13Q1) from 2001 to 2022 at a spatial resolution of 250 m and a temporal resolution of 16 days.
- Global gridded population data in 2022 from LandScan at a spatial resolution of 100 m.
- Classification of country income levels (Hamadeh et al. 2022).

- **Method**

This case study used the 2022 land-cover type data (IGBP classification) in the MCD12Q1 dataset to screen urban patches with a contiguous UBA larger than 50 km^2 (large cities, hereafter). This method can not only identify key cities in major economies, but also select important yet small-scale cities in African regions. Furthermore, it allows for merging and processing urban enclave patches according to administrative divisions of cities. There are 1,944 such cities worldwide. Further, using EVI data with a 250 m spatial resolution as a greenness indicator, we applied the Mann–Kendall method to evaluate the trend of the annual maximum EVI ($V_{\max}$) for each urban pixel of the 1,944 large cities from 2001 to 2022.

Then, we calculated the ratio (R_g) of the area of the pixels with a significantly increasing $V_{\max}$ trend ($p < 0.05$) to the UBA in 2022 of the corresponding city. Using this method, we evaluated urban greening change from 2001 to 2022.

Urban greening is an important part of the urban ecosystem. It not only has many benefits such as regulating rainwater, reducing air pollution, and reducing the urban heat island effect, but also reduces violent crimes and has a positive effect on the physical and mental health of urban residents (Giles-Corti et al. 2016). Therefore, this case study regarded the residents living in the UBA pixels with significant greening trends as the direct beneficiary population. Through the spatial overlay analysis of significant urban greening in UBAs and the global gridded population data (LandScan) in 2022, the population directly benefiting from urban greening can be estimated in each city.

- **Results and Analysis**

1. **China Has the Largest Urban Area of Significant Greening in the World**

Figure 5.25 shows the spatial distribution of R_g and the beneficiary population in 1,944 large

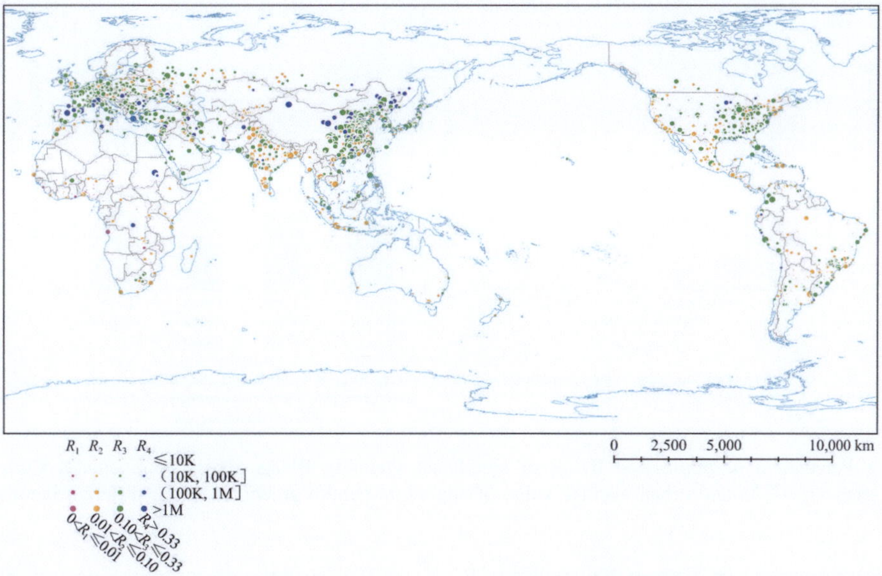

Fig. 5.25 Spatial distribution of R_g and the beneficiary population in 1,944 large cities from 2001 to 2022

cities around the world from 2001 to 2022. In the figure, each dot represents a city, with different colors indicating varying greening ratio levels (R_g): magenta for R_1 ($0 < R_1 \leq 0.01$), orange for R_2 ($0.01 < R_2 \leq 0.10$), green for R_3 ($0.10 < R_3 \leq 0.33$), and blue for R_4 ($R_4 > 0.33$). The size of the dots corresponds to the graded scale of the direct beneficiary population of significant urban greening in each city. Globally, the large cities with large values of R_g (indicated by blue and green points on the map) are located mainly in East Asia, Europe, and the Eastern United States, and sparsely in Africa, Oceania and South America. The significant greening degree of UBAs is closely related to a city's level of economic development. Taking China as an example, although it remains an upper-middle-income country, the extent of urban environmental improvement in China has attracted global attention due to strong policy support (e.g., the National Forest City Project) and continuous socioeconomic progress. The 340 selected Chinese cities accounted for only 17.5% of the global UBAs, but contributed 28% of significantly greening UBAs among the 1,944 large cities in the world in this case study. This shows that from 2001 to 2022, China's urban ecological environment improved significantly and had a comparative advantage globally. In contrast, in Africa, Latin America and other regions, many cities lack planning for development and sustained policy support. Not only have a large number of urban slums emerged in the process of urbanization, but cities also lack basic infrastructure, the urban ecological environment is becoming increasingly worrying.

2. **Nearly Half of the World's Population Benefiting from Significant Urban Greening is in China**

Globally, about 355 million urban dwellers live in significant greening UBAs. The proportions of the beneficiary population by continent are as follows: Asia (69.8%), Europe (11.8%), North America (7.5%), South America (6.3%), Africa (4.1%), and Oceania (0.5%) (Fig. 5.26a). The top five large cities with the largest beneficiary population in the world are the Pearl River Delta (16.24 million), the Yangtze River Delta (13.95 million), Tokyo Bay (9.27 million), Seoul (8.97 million) and Beijing (8.14 million).

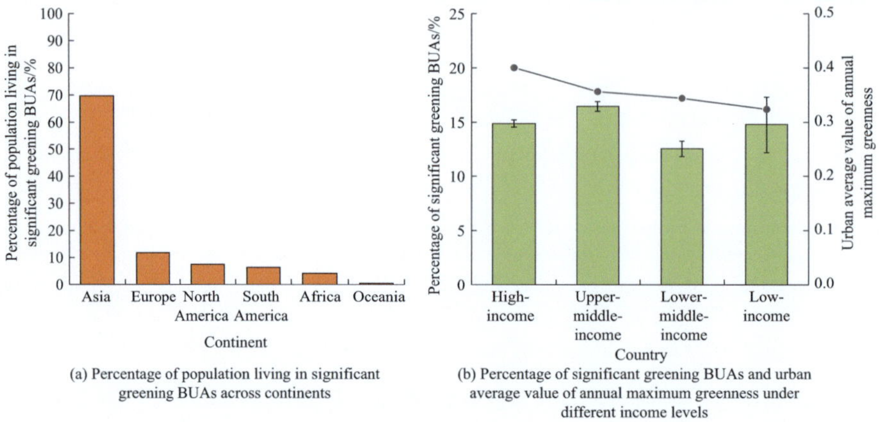

Fig. 5.26 **a** Percentage of population living in significant greening BUAs across continents **b** Comparison of significant greening BUAs and urban average value of annual maximum greenness under different income levels

3. **The Improvement in Urban Ecosystems is Closely Related to Income Level, with the Most Notable Improvement Seen in Upper-Middle-Income Countries**

The average values of R_g ranging from high to low for the income groups defined by the WB are 16.45% for upper-middle-income countries (UM), 14.88% for high-income countries (H), 14.80% for low-income countries (L), and 12.53% for lower-middle-income countries (LM) (Fig. 5.26b, bar graph). Using the average value of annual maximum greenness from 2019 to 2022 to represent the current environmental status of cities (EVI_{city}) (Fig. 5.26b, broken line graph), it was found that the EVI_{city} is the highest (0.40) in high-income countries, while it is 0.356, 0.344, and 0.324 in upper-middle-income, lower-middle-income, and low-income countries respectively, suggesting the need for a major improvement in their urban environments.

Highlights

- Based on the MODIS MOD13Q1 dataset (enhanced vegetation index data), MCD12Q1 land cover type dataset, LandScan global population grid data, and national income level data, we screened 1,944 large cities around the world with a UBA of more than 50 km^2. Then we evaluated the changes in greenness of UBAs and the direct beneficiary population in these cities from 2001 to 2022.
- With only 17.5% of the world's total UBAs, China accounted for 28% of the significant greening UBAs. China has made remarkable achievements in building an ecological civilization in cities.
- Globally, about 355 million urban dwellers live in significant greening UBAs, with 42% of the world's population benefiting from significant urban greening residing in China. The Pearl River Delta and the Yangtze River Delta are the top two urban agglomerations with the largest populations in the world benefiting from the significant greening of cities.

- **Discussion and Outlook**

This case study used the EVI of the MOD13Q1 dataset, land-cover change data, gridded population data, and national income level data to calculate the trends of annual maximum EVI and significance levels of UBAs in 1,944 large cities around the world from 2001 to 2022. This case study further analyzed the different ratios of greening UBAs of large cities at different income levels, and then evaluated the distribution of the direct beneficiary population in 2022.

Compared with the traditional scheme of calculating the changes of urban green space areas, the method used in this case study does not rely on high-resolution data, but rather monitors the overall greenness change of urban pixels at mesoscale resolution. The method detects changes in the greenness of urban pixels caused by newly added parks and green spaces on UBA pixels, and even detects changes in greenness caused by the natural growth of trees on streets.

In the context of global climate change, as a major contributor to urban carbon sinks, the significant increase in urban greening will help China achieve high-quality development. At the same time, for most developing countries, the results of this case study can provide data support, improve urban governance for these countries, and provide low-cost solutions for urban sustainable development monitoring in developing countries.

5.3.6 Spatiotemporal Trends of Atmospheric Particulate Matter in Global Typical Regions

Target: SDG11.6 By 2030, reduce the adverse per capita environmental impact of cities, including by paying special attention to air quality and municipal and other waste management.

- **Background**

With the continuous growth of the global population and the acceleration of urbanization, cities are constantly expanding in size and facing more and more environmental challenges. Among them, many cities are facing serious air pollution problems, and atmospheric particulate matter is one of the main pollution causes. Air pollutants have become the fourth leading risk factor for premature death globally, increasing the risk of premature death due to cardiovascular disease, respiratory disease, lung cancer, and lower respiratory tract infection. According to the 2022 World Air Quality Report, only 5% of countries comply with the World Health Organization (WHO)'s Global Air Quality Guidelines, with the most polluted regions mainly concentrated in Central Asia, South Asia, and Africa, while low- and middle-income countries have the highest levels. According to the WHO, about 99% of the global population breathes unhealthy air, and approximately 7 million people die each year from diseases related to air pollution. Low- and middle-income countries have the highest proportion of premature deaths due to air pollution, accounting for 91% of the total number of premature deaths related to outdoor air pollution worldwide.

The above data indicate that reducing global air pollution is of great significance for protecting human health and improving environmental quality. However, the monitoring capability of global atmospheric particulate matter varies by region and country, and is influenced by factors such as resources, technologies, monitoring network density, data sharing, and monitoring standards. Therefore, this case study analyzed and discussed the spatiotemporal trends of atmospheric particulate matter in typical global regions by integrating monitoring data of air pollutants from multiple countries and regions, conducting data screening, filtering, and quality control.

- **Data**

– Publicly available atmospheric station observation data from environmental protection departments in multiple countries and regions around the world from 2015 to 2022.

- **Method**

This case study used methods such as median filtering, low-pass filtering, cross comparison, and outlier detection to control the quality of data from different stations, thereby integrating years of global atmospheric pollutants and meteorological parameter datasets. Firstly, GIS mapping tools were used to depict the spatial distribution of various air pollutants. Then, the annual average concentration of global air pollutants was calculated through statistical analysis, and trend analysis was completed based on linear regression.

- **Results and Analysis**

From 2015 to 2022, the average concentration of PM2.5 was 20 μg/m^3, with 92% of urban sites exceeding the annual average concentration standard of 10 μg/m^3 set by the WHO in 2005. High PM2.5 concentrations were mainly found in densely populated areas of Asia, the Middle East, and Western Europe. In Asia, India and China had higher PM2.5 concentrations, with only a few cities meeting the WHO's air quality standards. In the Middle East, due to its arid and semi-arid climate, dust contributed significantly to total particulate matter, resulting in moderate concentrations. PM2.5 concentrations in Europe were relatively low, with slightly higher levels in Eastern Europe compared to Western Europe. North America had the lowest levels of PM2.5 concentrations among all continents, with higher concentrations observed mainly in major cities on the eastern and western coasts. Due to limited monitoring stations, available data for South America and Africa could not reflect the overall situation on those continents, but the data did indicate mild to moderate pollution levels in the relevant cities.

The spatial distribution of NO_2 was similar to that of PM2.5, while O_3 showed some differences from NO_2 and PM2.5. NO_2 was mostly found in heavily populated cities, but had lower concentrations in remote areas. O_3 remained high in Asia, mainly due to the continuous increase in volatile organic compound emissions in recent years. However, India, one of the regions with severe PM2.5 pollution, had relatively low O_3 concentrations, mainly attributable to the uptake of oxidants by particulate matter of high concentrations in the atmosphere.

From 2015 to 2020, PM2.5 concentrations showed an overall declining trend, with the global average concentration decreasing from 21.6 μg/m^3 to 17.2 μg/m^3. After 2020, there was a slight upward trend, reaching 19.4 μg/m^3 in 2022. The interannual variation trend of NO_2 was similar to that of PM2.5, with the global average concentration decreasing from 12.7 ppb in 2015 to 9.9 ppb in 2020 and then reaching a peak of 13.1 ppb in 2022. The interannual variation trend of O_3 showed slight differences from PM2.5 and NO_2. Except for a low value in 2015, the average concentration remained around 23.6 ppb from 2016 to 2018 and then decreased significantly over the next four years, stabilizing at around 21.2 ppb (Fig. 5.27). The main pollutants in the atmosphere reached their lowest levels during the study period in 2020 and 2021, primarily due to the slowdown in global economic development and reduced human activities during the COVID-19 pandemic, which led to a decrease in atmospheric pollutant emissions.

Highlights

- Using observation data from environmental protection departments in multiple countries and regions around the world, we integrated and produced the global atmospheric particulate matter station monitoring dataset from 2015 to 2022.
- The long-term trends of global atmospheric pollutants were analyzed, and the main reasons for these trends based on geographical regions, population, economy, and other factors were identified.

- **Discussion and Outlook**

This case study is based on observation data from environmental protection departments in multiple countries and regions around the world, and uses data cleaning methods to produce the global atmospheric particulate matter station monitoring dataset. This dataset provides global monitoring data on atmospheric particulate matter and can be used to analyze the spatiotemporal variation trends of global atmospheric particulate matter. Through analysis, the changes in global atmospheric particulate matter emissions can be revealed, and the impact of global events on atmospheric environmental quality can be studied. This helps management departments and the public better understand the severity and changing trends of global air pollution problems, providing a scientific basis for formulating environmental policies and taking corresponding measures.

In the future, this dataset will be further improved to refine the long-term variation trends

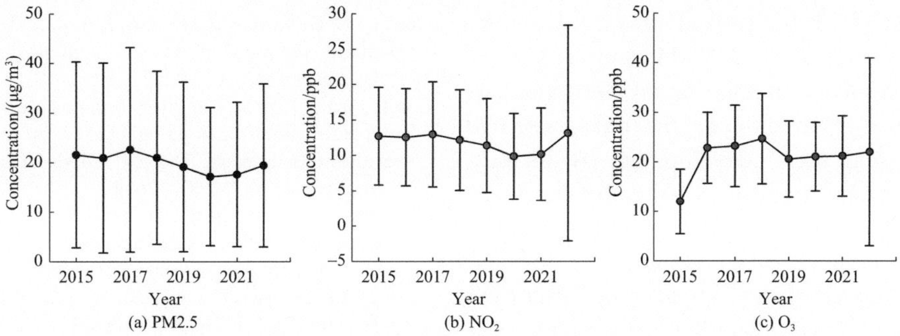

Fig. 5.27 Interannual variation trends in global PM2.5, NO_2 and O_3 concentrations during 2015–2022. *Note* 1 ppb = 10^{-9}

of atmospheric pollutants in different geographical regions. This will help to more accurately assess the atmospheric environmental quality in different regions and implement more targeted environmental management and improvement measures. In addition, by incorporating various parameters such as meteorology, population, geography, and economy, attribution analysis tools can reveal the main sources and influencing factors of air pollution, providing more specific guidance strategies and solutions for reducing air pollution.

5.4 Summary

This chapter explores progress in sustainable cities and communities, focusing on urban public transport, urbanization, heritage preservation, urban disaster management, urban air pollution, and open public spaces in cities. Research findings reveal that: The proportion of the population with convenient access to public transport has increased in major global cities; Cities in the Global South Countries have developed rapidly over the past decade, with China's urban area growth rate being nearly double that of Australia and the United States; Global heritage preservation shows positive development trends; Since the implementation of the Sendai Framework, regional differences are evident in annual average mortality and direct economic losses caused by extreme disasters; Developing countries have shown the most significant increase in green spaces within UBA globally; Multiple factors influence the sources of atmospheric particulate matter in typical global cities. These findings provide critical support for detailed analysis of regional indicator progress and evidence-based decision-making for sustainable cities and communities.

Based on the research in this chapter, we propose the following recommendations: (1) To increase the proportion of the population with convenient access to urban public transport and address the issue of uneven public transport accessibility globally, efforts should be made to strengthen the construction and improvement of urban public transport systems and promote the integration of public transport planning with urban planning. (2) Given the dynamic nature of urban development, global urban boundary datasets should be regularly updated to support future research in various urban domains, such as the impacts of urbanization on urban ecosystems, urban landscapes, and urban climate. Moreover, although cities in the Global South Countries are developing rapidly, the fast expansion of urban land poses significant threats to urban environments and ecosystems. Therefore, in the future, financial and technical support should be strengthened for developing countries, providing effective guidance, along with more scientific direction and policy recommendations to promote the sustainable development of cities and communities. (3) Data integrity and accuracy are crucial for heritage preservation. The WDPA should be updated promptly, and information on the WHC website should be improved to ensure precise data on the boundaries of World Natural Heritage Sites and Mixed Cultural-Natural Heritage Sites worldwide. (4) To mitigate the risks posed to economic and social development by increasingly severe and frequent extreme climate events under climate change, it is necessary to further strengthen research on the patterns and mechanisms of climate disasters, improve the forecasting and early warning capabilities for frequent and high-risk climate disasters, and enhance the monitoring and risk assessment of extreme climate disasters. (5) In comprehensive efforts to improve global air quality and conduct attribution studies, attention should be paid not only to areas with existing well-established monitoring networks and complete data, but also to regions with inaccurate or incomplete data, such as the Gobi Desert region. (6) Middle- and low-income countries should increase the provision of urban green infrastructure to address the challenges facing urban sustainable development.

Looking forward, with the continuous development of Big Earth Data technology, we will be able to monitor and evaluate urban development trends more comprehensively and accurately. This will allow us to make more effective use of information on population flow and resource utilization, provide rich data for sustainable cities and communities, inform policy on new indicator development, and offer a scientific basis for government planning and decision-making.

References

Chen Y, Bouferguene A, Shen YH et al (2019) Assessing accessibility-based service effectiveness (ABSEV) and social equity for urban bus transit: a sustainability perspective [J]. Sustain Cities Soc 44:499–510

Chen MX, Chen LK, Cheng JF et al (2022) Identifying interlinkages between urbanization and Sustainable Development Goals [J]. Geogr Sustain 3(4):339–346

Elvidge CD, Tuttle BT, Sutton PC et al (2007) Global distribution and density of constructed impervious surfaces [J]. Sensors 7(9):1962–1979

Esch T, Marconcini M, Felbier A et al (2013) Urban footprint processor: fully automated processing chain generating settlement masks from global data of the TanDEM-X mission [J]. IEEE Geosci Remote Sens Lett 10(6):1617–1621

Giles-Corti B, Vernez-Moudon A, Reis R et al (2016) City planning and population health: a global challenge [J]. The Lancet 388(10062):2912–2924

Gong P, Wang J, Yu L et al (2013) Finer resolution observation and monitoring of global land cover: first mapping results with Landsat TM and ETM+ data [J]. Int J Remote Sens 34(7):2607–2654

Hamadeh N, van Rompaey C, Metreau E et al (2022) New World Bank country classifications by income level: 2022–2023 [R]. https://www.blogs.worldbank.org/en/opendata/new-world-bank-country-classifications-income-level-2022-2023 [2024-10-09]

Homer C, Dewitz J, Yang LM et al (2015) Completion of the 2011 national land cover database for the conterminous United States–representing a decade of land cover change information [J]. Photogramm Eng Remote Sens 81(5):345–354

Hu SG, Tong LY, Frazier AE et al (2015) Urban boundary extraction and sprawl analysis using Landsat images: a case study in Wuhan, China [J]. Habitat Int 47:183–195

IPCC. 2023. Climate Change 2023: Synthesis Report [R]. https://www.ipcc.ch/report/ar6/syr/downloads/report/IPCC_AR6_SYR_LongerReport.pdf [2024-02-20]

IQAir (2023) World air quality report 2022 [R]. https://www.iqair.com/in-en/newsroom/world-air-quality-report-press-release-2022 [2024-02-20]

Kocabas V, Dragicevic S (2006) Assessing cellular automata model behaviour using a sensitivity analysis approach [J]. Comput Environ Urban Syst 30(6):921–953

Li XC, Zhou YY, Zhu ZY et al (2018) Mapping annual urban dynamics (1985–2015) using time series of Landsat data [J]. Remote Sens Environ 216:674–683

Liang X, Liu XP, Li X et al (2018) Delineating multi-scenario urban growth boundaries with a CA-based FLUS model and morphological method [J]. Landsc Urban Plan 177:47–63

Mbow H-OP et al (2017) Special Report on climate change, desertification, land degradation, sustainable land management, food security, and greenhouse gas fluxes in terrestrial ecosystems (SR2) [R]. https://www.ipcc.ch/site/assets/uploads/2018/07/sr2_background_report_final.pdf [2024-02-20]

Mittal S, Dai HC, Shukla PR (2016) Low carbon urban transport scenarios for China and India: a comparative assessment [J]. Transp Res Part D: Transp Environ 44:266–276

Peng J, Zhao SQ, Liu YX et al (2016) Identifying the urban-rural fringe using wavelet transform and kernel density estimation: a case study in Beijing City, China [J]. Environ Model Softw 83:286–302

Ravensbergen L, Van Liefferinge M, Isabella J et al (2022) Accessibility by public transport for older adults: a systematic review [J]. J Transp Geogr 103:103408

Saghapour T, Moridpour S, Thompson RG (2016) Public transport accessibility in metropolitan areas: a new approach incorporating population density [J]. J Transp Geogr 54:273–285

Saif MA, Zefreh MM, Torok A (2018) Public transport accessibility: a literature review [J]. Period Polytech Transp Eng 47(1):36–43
Sun LQ, Chen J, Li QL et al (2020) Dramatic uneven urbanization of large cities throughout the world in recent decades [J]. Nat Commun 11(1):5366
Takase C (2018) Implementing SDG 11-key elements, challenges and opportunities. module 4: SDGs-sustainable cities and communities 2018 executive training course for policymakers on the 2030 agenda and the sustainable development goals (SDGs)[EB/OL]. https://sustainabledevelopment.un.org/content/unosd/documents/4057Module%204%20SDG%2011_Chicako%20Takase.pdf [2024-10-16]
Tian WM (2012) Calculation and trend analysis of Gini coefficient in China [J]. J Humanities 2:56–61 (In Chinese)
Tiwari G, Phillip C (2021) Development of public transport systems in small cities: a roadmap for achieving sustainable development goal indicator 11.2 [J]. IATSS Res 45(1):31–38
UNDRR (2023) The report of the midterm review of the implementation of the Sendai Framework for disaster risk reduction 2015–2030 [R]. https://www.undrr.org/publication/report-midterm-review-implementation-sendai-framework-disaster-risk-reduction-2015-2030 [2024-02-20]
UNESCO & World Heritage Center (2023) The operational guidelines for the implementation of the world heritage convention—2023 operational guidelines [EB/OL]. https://www.whc.unesco.org/en/guidelines [2024-02-20]
UN-Habitat (2022) World Cities Report 2022: Envisaging the Future of Cities [R]. https://unhabitat.org/sites/default/files/2022/06/wcr_2022.pdf [2024-02-20]
United Nations, Department of Economic and Social Affairs, Population Division (2019) World urbanization prospects: the 2018 Revision (ST/ESA/SER.A/420) [R]. https://www.un.org/development/desa/pd/sites/www.un.org.development.desa.pd/files/files/documents/2020/Jan/un_2018_wup_report.pdf [2024-02-20]
Wang L, Li CC, Ying Q et al (2012) China's urban expansion from 1990 to 2010 determined with satellite remote sensing [J]. Chin Sci Bull 57(22):2802–2812
Wang QF, Yu GR, He HL et al (2015) Thinking of construction of the nature reserve system and integrated management system in China [J]. Resour Sci 37(7):1357–1366 (In Chinese)
Xi FM, Davis SJ, Ciais P et al (2016) Substantial global carbon uptake by cement carbonation [J]. Nat Geosci 9(12):880–883
Zhang HC, Ning XG, Shao ZF et al (2019) Spatiotemporal pattern analysis of China's cities based on high-resolution imagery from 2000 to 2015 [J]. ISPRS Int J Geo Inf 8(5):241
Zhao M, Zhou YY, Li XC et al (2020) Building a series of consistent night-time light data (1992–2018) in Southeast Asia by integrating DMSP-OLS and NPP-VIIRS [J]. IEEE Trans Geosci Remote Sens 58(3):1843–1856
Zhou YY, Li XC, Asrar GR et al (2018) A global record of annual urban dynamics (1992–2013) from nighttime lights [J]. Remote Sens Environ 219:206–220

SDG 13 Climate Action

6

6.1 Background

Global greenhouse gas concentrations continue to rise, leading to record-breaking global average temperatures. This trend results in an increase in extreme weather events and an acceleration of glacier and ice cap melting, causing sea level rise and posing a threat to biodiversity (WMO 2023). The Global Risks Report 2023 indicates that the three greatest risks facing humanity in the next decade are the inability to mitigate climate change, the inability to adapt to climate change, and extreme natural disasters, all of which are closely related to climate change. Climate change is already seriously threatening human survival and development (WEF 2023). The UN *Sustainable Development Goals Report 2022* points out that the world is on the edge of climate catastrophes, and the window for avoiding it is closing. Global greenhouse gas emissions are still increasing, and all countries need to take immediate action to shift from the tipping point of climate disaster to a turning point towards sustainable future (United Nations 2022). After announcing carbon peaking and carbon neutrality goals in 2020, China released the *National Climate Change Adaptation Strategy 2035* in 2022, promoting enhanced adaptive capacity and proactive responses to climate change.

SDG 13 (Climate Action) aims to take urgent action to combat climate change and its impacts. However, it is currently the most data-deficient among all 17 SDGs, with only about 20% of countries having relevant data (United Nations 2022), and data with spatial and temporal information is even scarcer. Therefore, there is an urgent global need for data that can reflect overall progress towards SDG indicators while providing spatial details and temporal trends to inform decisions on disaster response and climate change mitigation.

The reports from 2019 to 2022 focused on five indicators: disaster losses (SDG 13.1.1), national disaster risk reduction strategies (SDG 13.1.2), proportion of local governments with disaster risk reduction strategies (SDG 13.1.3), greenhouse gas emissions (SDG 13.2.2), and climate change education (SDG 13.3.1). These reports provided indicator calculation methods, spatiotemporal data products, and scientific decision support at both the Chinese and global scales.

This year, we have reviewed the new progress in global greenhouse gas balance in response to SDG 13.2.2, providing scientific data support for formulating greenhouse gas emission reduction strategies.

6.2 Main Contributions

This chapter focuses on the specific goals of SDG 13.2, based on multi-source data such as satellite remote sensing, surveys, and statistics. Through

H. Guo, *Big Earth Data in Support of the Sustainable Development Goals (2023)—The Belt and Road*, Sustainable Development Goals Series, https://doi.org/10.1007/978-981-95-3178-3_6

Table 6.1 Cases and their main contributions

Target	Tiers	Cases	Contributions
SDG 13.2 Integrate climate change measures into national policies, strategies and planning	Tier I	Global Near-Real-Time Carbon Data	Data product: Daily carbon emission dataset of major countries and sectors worldwide from 2019 to 2022
	Tier I	Global fluxes of three greenhouse gases (CO_2, CH_4, and N_2O) from sources and sinks in terrestrial ecosystems	Decision support: Presenting a panoramic map of public awareness of climate change issues in seven major geographical regions of China

two cases, it provides a model method for monitoring and analyzing indicators for SDG 13, as well as data products with rich spatiotemporal information. Based on this, decision recommendations are proposed (Table 6.1).

6.3 Case Study

6.3.1 Global Near-Real-Time Carbon Data

Target: SDG 13.2 Integrate climate change measures into national policies, strategies and planning.

- **Background**

The main objective of SDG 13 (Climate Action) is to take urgent action to combat climate change and its impacts. It is one of the key goals of the SDGs (United Nations 2015). Phenomena such as global warming pose significant threats to human well-being and the health of the planet's ecosystems. Therefore, addressing extreme climate events, ecosystem impacts, and climate-related risks is crucial for mitigating economic, social, and environmental losses, and is vital for achieving sustainable development in countries around the world.

In response to the urgent need to combat climate change, the 21st session of the Conference of the Parties (COP 21) to the United Nations Framework Convention on Climate Change (UNFCCC) adopted the Paris Agreement, with the goal of limiting global temperature rise to well below 2 °C above pre-industrial levels, aiming for 1.5 °C. To achieve this, countries demonstrated strong ambitions and determination in climate governance by presenting their national climate change mitigation plans and carbon neutrality timelines. However, assessing the formulation and implementation of these mitigation measures requires more scientific, timely, and accurate high-resolution data.

Currently, the UN Inter-Agency and Expert Group on SDG Indicators (IAEG-SDGs) has established a monitoring indicator system (framework) for SDG 13.2, which includes annual greenhouse gas emissions as a key climate indicator. However, this indicator is insufficient to support the global carbon reduction and carbon neutrality goals. There remains a significant data gap in terms of geographic coverage and timeliness. National carbon emission data is currently based on annual national statistics, which suffer from a time lag of over one year. Moreover, global carbon emissions may have undergone dramatic changes due to unforeseen events such as the COVID-19 pandemic and other global crises. Therefore, establishing near-real-time global carbon emission indicators is crucial for timely, accurate, and reliable carbon emission monitoring and for evaluating SDG 13.

Since 2020, the Carbon Monitor team, led by Tsinghua University, has developed a near-real-time global carbon emission database, which presents the carbon emission dynamics of various countries in a scientific and transparent manner. In the context of the COVID-19 pandemic, the Carbon Monitor team continuously tracked changes in carbon emissions across countries, considering multiple policy objectives such as pandemic control, economic recovery, and carbon emission reduction. This achievement provides real-time, accurate carbon emission information to governments, research institutions, and businesses, facilitating more effective climate action planning and contributing positively to achieving SDG 13.

- **Data**

- Global atmospheric reanalysis dataset from 2019 to 2022, provided by the Copernicus Climate Change Service (C3S), with a spatial resolution of 0.25°.
- 2018 global carbon emission data, sourced from the Emissions Database for Global Atmospheric Research (EDGAR).
- 2019-2022 global national sectoral (electricity, industry, aviation, ground transportation, residential consumption, shipping) activity data, including energy consumption, industrial activity, global positioning system (GPS) data, and radar data, from national statistical agencies, open-source websites, and operating companies.
- 2019-2022 TROPOspheric Monitoring Instrument (TROPOMI) NO_2 column concentration data from ESA.

- **Method**

Countries are categorized by their carbon emission sources into six major sectors: electricity, industry, residential consumption, ground transportation, aviation, and shipping. Based on multi-source data such as statistical data, satellite remote sensing, and observational data, a near-real-time quantification method for carbon emissions was constructed, assessing the spatiotemporal dynamics of global carbon emissions from 2019 to 2022. This method utilized the linear relationship between carbon emissions and energy activity intensity (such as the relationship between emissions and fuel consumption, electricity generation, driving distance, etc.) to estimate carbon emissions in near-real-time within an uncertain range. The resulting dataset consisted of daily sector-specific carbon emission data for major countries from 2019 to 2022. By constraining the total daily carbon emissions at the national scale with these datasets and using NO_2 satellite observation concentration values for spatial distribution corrections, near-real-time daily global carbon emission data was produced. These data were then used to assess SDG 13 indicators and the progress of climate change mitigation from 2019 to 2022.

- **Results and Analysis**

The global and major emitting countries'/regions' daily carbon emissions from 2019 to 2022 are shown in Fig. 6.1. Daily global carbon emission patterns indicate higher emissions in winter and summer, and lower emissions in spring and autumn, forming a trend similar to that of a galloping horse. The highest global carbon emissions occur in winter (December), and the lowest in spring (April), with emissions rising slightly in summer and increasing rapidly in autumn. This pattern is linked to the concentration of populations in the Northern Hemisphere, where heating and cooling demands are higher in winter and summer. While seasonal variations in monthly emissions differ among countries due to geographic location, the daily emission trends in major carbon-emitting countries still follow the "galloping horse" curve. Additionally, global carbon data reflect a significant drop in anthropogenic carbon emissions in 2020 due to the COVID-19 pandemic, with a 5.4% decrease compared to 2019 (1.9 Gt CO_2), and a rebound in 2021 to near pre-pandemic levels, with an increase of 6.3% (2.1 Gt CO_2).

The global near-real-time carbon emission map (Fig. 6.2) covers over 90% of the world's countries and regions, with a spatial resolution of $0.1° \times 0.1°$, providing insights into global carbon emission hotspots and revealing emissions patterns at annual, seasonal, monthly, and weekly scales. Major carbon emission hotspots are concentrated in Eastern North America, Western Europe, South Asia, and Eastern China. High-resolution carbon emission maps enable analysis at finer spatial scales, allowing the study of carbon emission trends at provincial and city levels.

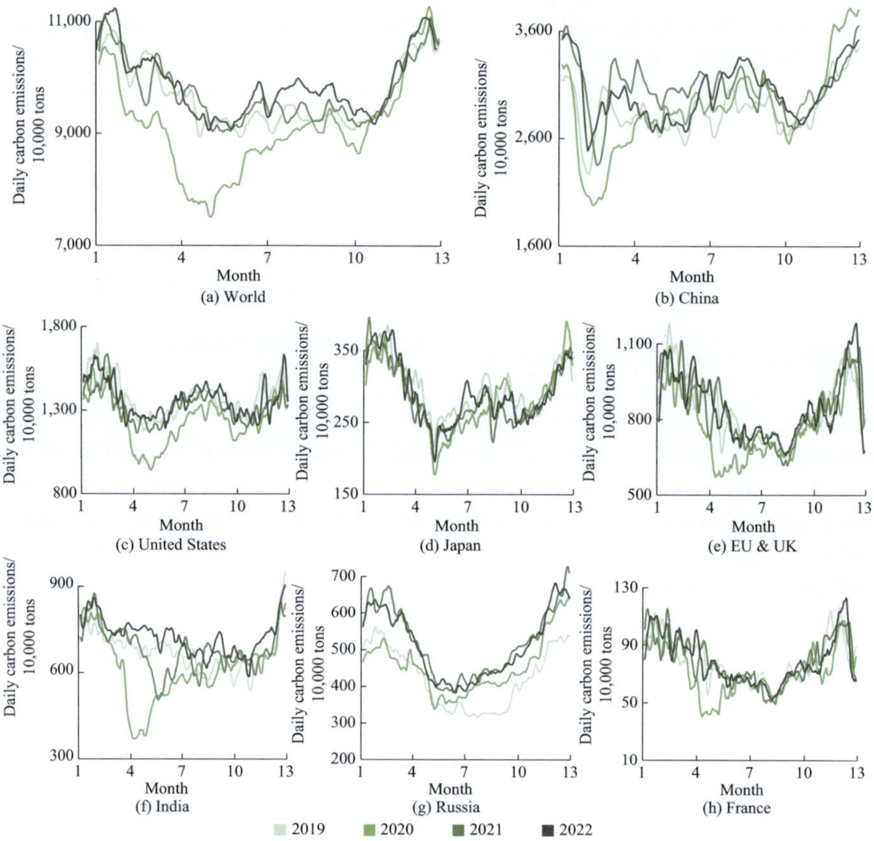

Fig. 6.1 Daily carbon emission data of global and major emitting countries/regions (2019–2022)

Highlights

- Based on the UNFCCC and multi-source near-real-time big data, the global national-scale near-real-time carbon emission monitoring and evaluation system was developed, providing timely and accurate reflections of carbon emission responses to emergencies and the latest progress in climate change mitigation actions.
- Global CO_2 emissions in 2022 increased by 1.5% compared to 2021 (7.9% compared to 2020 and 2.0% compared to 2019), reaching 36.1 Gt CO_2. If global emissions remain at the 2022 level, the remaining carbon budget under the 1.5 °C target will be exhausted within the next 2-7 years (with a 67% probability).
- The daily carbon emission dynamics of the world and major emitting countries/regions show higher emissions in winter and summer and lower emissions in spring and autumn. This pattern is associated with the high population density in the Northern Hemisphere and the heating and cooling demands in winter and summer.

- **Discussion and Outlook**

Under the SDG 13 monitoring framework, this case study proposes a carbon emission quantification approach based on multi-source activity data, addressing issues such as the time lag, uncertainty, and data accuracy in the current greenhouse gas emission accounting system. This innovative approach enables near-real-time dynamic monitoring of carbon emissions and the construction of a consistent, comparable global dataset of daily

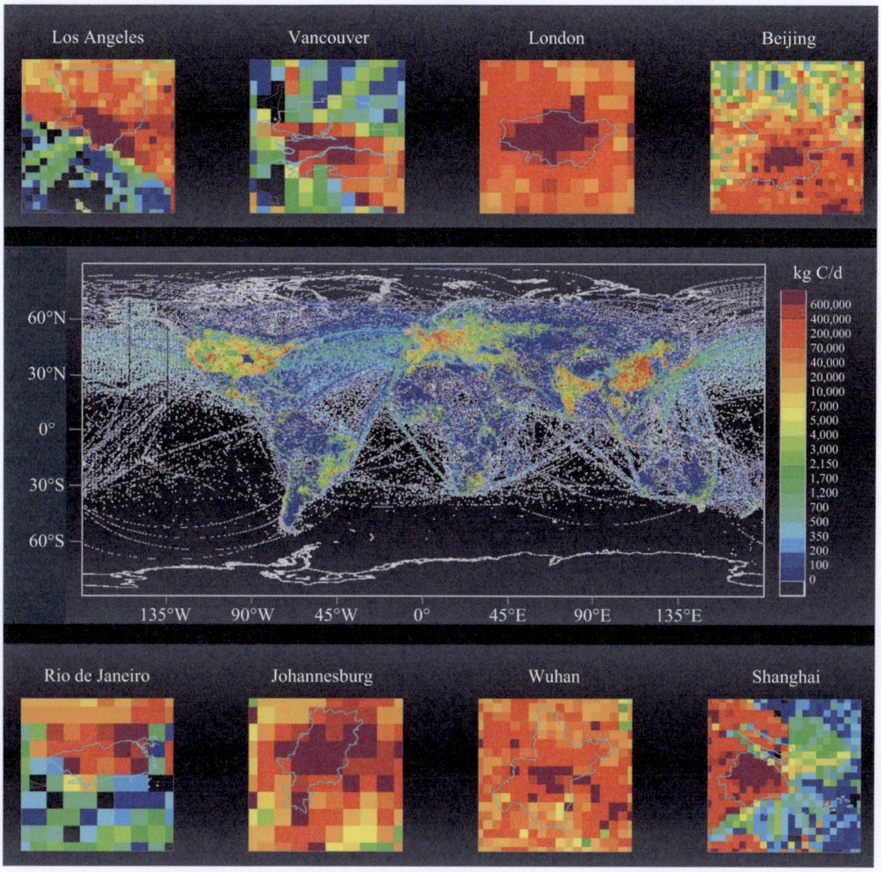

Fig. 6.2 Global near-real-time carbon emission map at the 0.1° × 0.1° daily scale and emission trends for major cities

sector-specific carbon emissions for major countries, which helps evaluate global carbon reduction and carbon neutrality progress. This approach aids in taking more effective low-carbon actions, providing an important data foundation for optimizing carbon reduction policies and responding to emergencies such as extreme weather, natural disasters, and public crises. It also offers vital data support for evaluating SDG 13 indicators.

Global near-real-time carbon data shows that, although the COVID-19 pandemic led to a record decline in global anthropogenic carbon emissions in 2020, global carbon emissions have since returned to pre-pandemic levels. Despite efforts to reduce fossil fuel use, global carbon emissions continue to rise. If emissions remain at the 2022 level, the remaining carbon budget will further shrink. Therefore, countries must act immediately to fulfill net-zero emission commitments and enhance near-real-time greenhouse gas monitoring capabilities to assess the effectiveness of mitigation measures and climate change mitigation progress.

6.3.2 Global Fluxes of Three Greenhouse Gases (CO_2, CH_4, and N_2O) from Sources and Sinks in Terrestrial Ecosystems

Target: SDG 13.2 Integrate climate change measures into national policies, strategies and planning.

- **Background**

SDG 13.2, "Integrate climate change measures into national policies, strategies and planning", is one of the important objectives of the SDGs (United Nations 2015), which is directly related to whether human society can take timely and effective measures to curb the continuous increase of greenhouse gas concentrations in the atmosphere, effectively respond to climate change, and maintain the sustainable development of human economic and social systems, industrial and agricultural production, and natural ecosystems.

One of the most important foundations for incorporating initiatives to address climate change into national policies and planning is the accurate quantification of greenhouse gas fluxes from sources and sinks, which is the basis for the formulation of national greenhouse gas emission reduction policies. In this regard, as early as 1996, the IPCC formulated the Guidelines for National Greenhouse Gas Inventories. The guidelines provide different tiers of accounting for greenhouse gas fluxes by sources and sinks in a variety of sectors, including energy, industry, forestry, agriculture and land use change, and waste. Among these, the quantification of greenhouse gas fluxes in terrestrial ecosystems presents the greatest uncertainty. The highest-tier method recommended in the guidelines involves using terrestrial ecosystem models for national-scale inventories. Particularly on a global scale, conducting long-term assessments of the net fluxes of the three greenhouse gases (CO_2, CH_4, and N_2O) in terrestrial ecosystems is critical for effectively managing terrestrial ecosystems, enhancing their carbon sink function, and scientifically formulating climate change mitigation policies.

This case study utilizes the integrated biosphere simulator (IBIS) model, combining with Big Earth data, to simulate the sources and sinks of the three greenhouse gases (CO_2, CH_4, and N_2O) in global terrestrial ecosystems. By analyzing the spatiotemporal patterns of greenhouse gas fluxes in terrestrial ecosystems, this case study evaluates the sources and sinks in China's terrestrial ecosystems and their contributions to global fluxes, and quantifies the impacts of climate change and human activities on these fluxes. It also identifies the key factors driving the changes. These research results provide a solid scientific basis for the preparation of national communications and biennial update reports under SDG 13.2.

- **Data**

- 1980-2022 global atmospheric reanalysis dataset, provided by the ECMWF, with a spatial resolution of 9 km.
- 1980-2022 global land cover dataset (Land-Use Harmonization, LUH2), provided by the University of Maryland, with a spatial resolution of 25 km.
- 2000-2018 global wetland distribution dataset (Wetland Area and Dynamics for Methane Modeling, WAD2M), provided by the University of Maryland, with a spatial resolution of 25 km.
- 1980-2015 China Forest Cover Dynamics (CFCD) dataset, provided by Sun Yat-sen University, with a spatial resolution of 1 km.
- 2000-2020 Asian monsoon region rice distribution dataset, provided by Beijing Normal University, with a spatial resolution of 1 km.

- **Method**

This case study employs the IBIS model (Yuan et al. 2014) to calculate the sources and sinks of the three greenhouse gases (CO_2, CH_4, and N_2O) in global terrestrial ecosystems. The model fully integrates biological, chemical, and physical processes in terrestrial ecosystems, making it one of the few models worldwide capable of simultaneously simulating the sources and sinks of all three greenhouse gases. Since 2019, the IBIS model has been part of the Global Carbon Project, providing annual global terrestrial ecosystem CO_2 flux data directly used in IPCC assessment reports. Building on the IBIS model, this case study incorporates the latest Big Earth Data products, including land-use change and vegetation type distribution, to improve the accuracy of global terrestrial ecosystem greenhouse gas flux simulations. Notably, this case study utilizes datasets specifically designed for China to accurately capture the spatiotemporal variations in greenhouse gas fluxes from sources and sinks and analyze China's contributions to global terrestrial ecosystem fluxes of the three greenhouse gases.

- **Results and Analysis**

This case study first presents the sources and sinks of the three greenhouse gases in global terrestrial ecosystems. For CO_2, the global terrestrial ecosystem showed a significant carbon sink with an average emission intensity of 2.36 Pg C/a (Fig. 6.3). However, for CH_4 and N_2O, the global terrestrial ecosystems were emission sources, with average emission intensities of 173 Tg C/a and 8.9 Tg N/a from 1980 to 2022, respectively. The long-term trends of these three greenhouse gases also showed notable differences. The carbon sink increased significantly from 1980 to 2000, with an annual growth rate of 0.17–0.19 Pg C/a (Fig. 6.4a). However, since 2000, the growth of the global terrestrial ecosystem carbon sink has stagnated and even shown a slight declining trend. CH_4 emissions exhibited a pattern of initial increase followed by a decrease, with the turning point occurring around 2012 (Fig. 6.4b). N_2O emissions increased significantly from 1980 to 2016 but began to decline after 2016 (Fig. 6.4c).

China's terrestrial ecosystem greenhouse gas emissions also showed significant long-term trends. Between 1980 and 2022, the average carbon sink strength of China's terrestrial ecosystems was 0.27 Pg C/a, accounting for 11% of the global carbon sink. Notably, thanks to the large-scale ecological projects implemented by China, its terrestrial carbon sink has increased more than the global carbon sink, and thus its contribution to the global carbon sink has also shown a significant increase. The average emission intensity of China's terrestrial CH_4 was 11.40 Tg C/a, and the emission has shown a decreasing and then increasing trend since 1980, which is consistent with the change in China's rice cultivation area. China's terrestrial ecosystems contributed 6.5% of global terrestrial CH_4 emissions, with this contribution also following a pattern of decline and subsequent increase. The average intensity of terrestrial N_2O emissions in China was 1.25 Tg N/a. Since 2016, environmental policies implemented by the Chinese government to reduce fertilizer use in agriculture have led to a significant decline in N_2O emissions. Between 2010 and 2022, China's terrestrial ecosystems contributed 14.66% of global N_2O emissions (Fig. 6.5b).

Highlights

- Based on the IBIS model and Big Earth Data, this case study comprehensively assessed the intensity and long-term trends of the sources and sinks of the three greenhouse gases (CO_2, CH_4, and N_2O) in global terrestrial ecosystems, with a focus on quantifying China's contributions to global fluxes.
- Compared to the global average, the carbon sink in China's terrestrial ecosystems exhibited a more significant increasing trend, leading to a notable rise in its contribution to the global terrestrial carbon sink since 1980.
- The CH_4 emissions from global terrestrial ecosystems showed a trend of initial decline followed by an increase since 1980, and China's terrestrial CH_4 fluxes exhibited a similar pattern.
- The N_2O emissions from China's terrestrial ecosystems began to show a significant decline starting in 2016, which also contributed to a gradual reduction in global N_2O fluxes.

- **Discussion and Outlook**

This case study, based on the IBIS model and the latest Big Earth Data, simulated the sources and sinks of the three greenhouse gases (CO_2, CH_4, and N_2O) in global terrestrial ecosystems, with a particular focus on China's contributions to global fluxes and their changing trends. It lays a methodological foundation for comprehensive assessments of terrestrial ecosystem greenhouse gas fluxes at national and global scales. More importantly, according to the Paris Agreement,

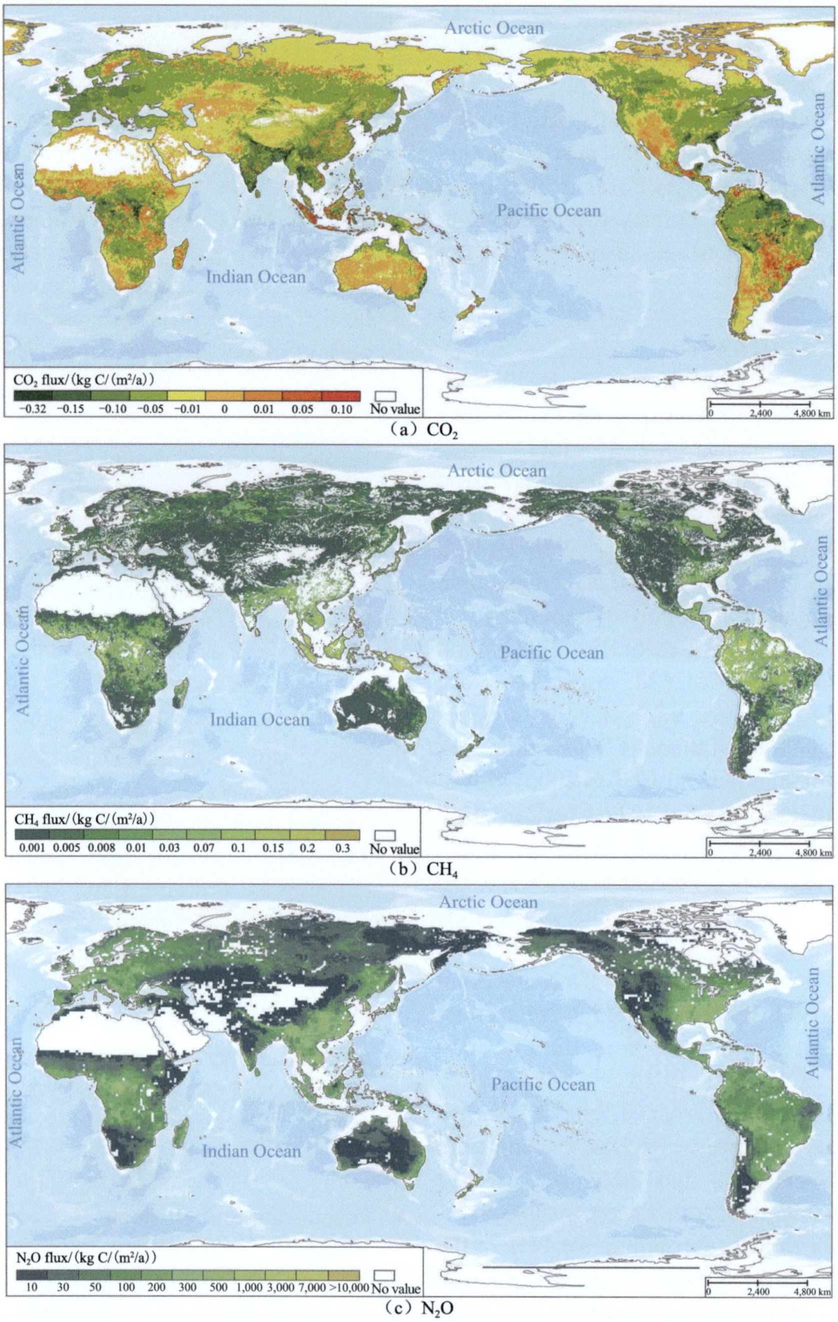

Fig. 6.3 Global average of three greenhouse gas fluxes from sinks in terrestrial ecosystems from 2010 to 2022

starting in 2024, all parties are required to submit national greenhouse gas inventories every two years. This means that developing countries, including China, will need to report their greenhouse gas emissions regularly, just as developed countries do. The accounting of global terrestrial ecosystem greenhouse gas fluxes of the three greenhouse gases in this case study provides an independent reference for other countries' emission inventories. Furthermore, the methodology

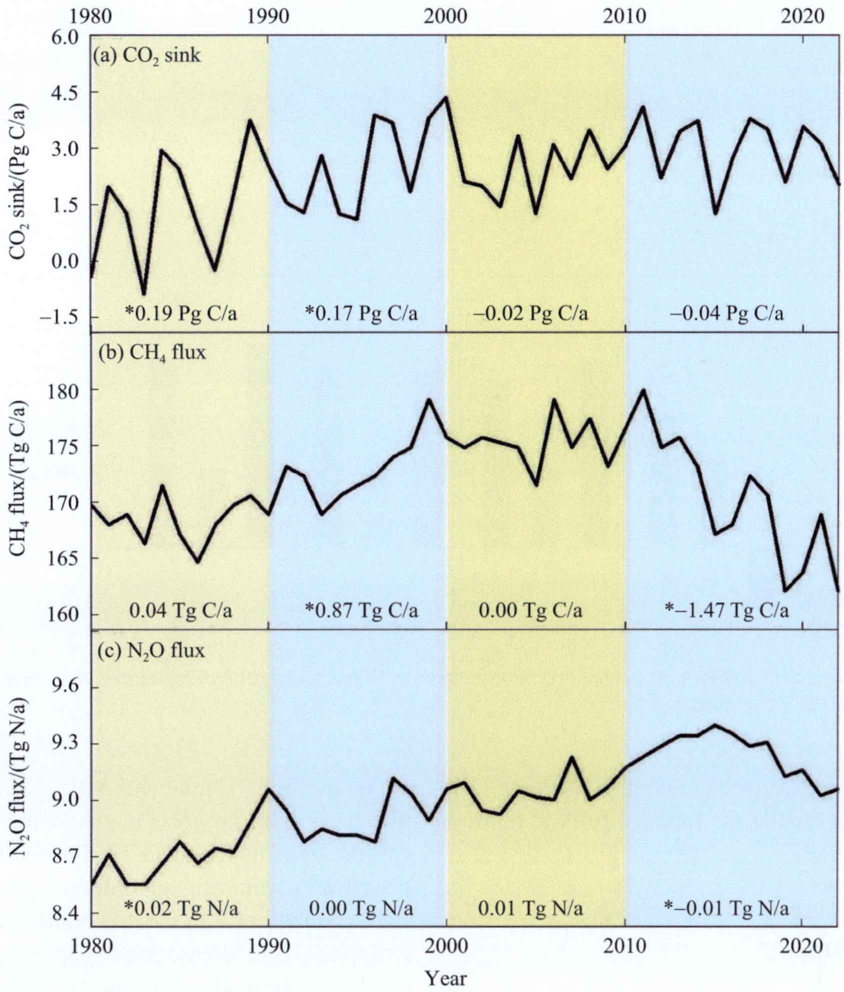

Fig. 6.4 Long-term trends in the fluxes of three greenhouse gas sources and sinks in global terrestrial ecosystems. *Note* The numbers in the figure indicate the trend over the time period, where * indicates that the trend is statistically significant ($p < 0.05$)

used in this case study can directly serve as a foundational approach for developing countries to compile national greenhouse gas inventories, which is highly significant for achieving SDG 13.2.

At the same time, the changes in China's contributions to global greenhouse gas fluxes demonstrated in this case study highlight China's outstanding efforts in addressing climate change. On the one hand, large-scale ecological projects implemented by the Chinese government since the early 1980s have significantly increased China's terrestrial carbon sinks. This increase has exceeded the global average, thereby significantly enhancing China's contribution to global carbon sinks. On the other hand, since 2016, the Chinese government has significantly reduced the use of chemical fertilizers in agriculture, resulting in a marked decrease in terrestrial N_2O emissions, which has played a critical role in reducing global emissions. However, it should be noted that the recent expansion of rice cultivation areas has led to a significant increase in CH_4 emissions from China's terrestrial ecosystems. Therefore,

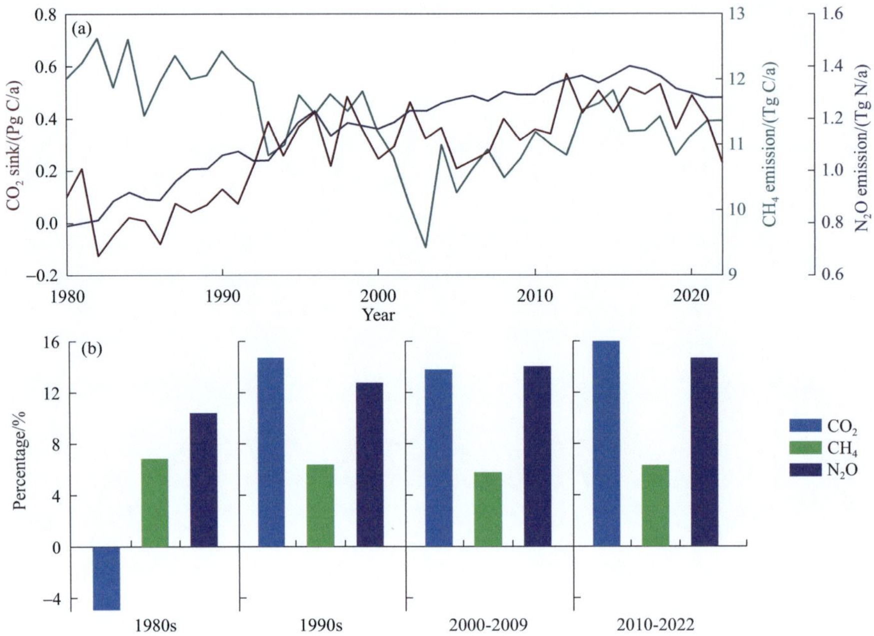

Fig. 6.5 **a** Sources and sinks of the three greenhouse gases in China's terrestrial ecosystems and **b** their proportional contribution to the global total

strengthening rice management to mitigate CH_4 emissions is crucial for future efforts to address climate change.

6.4 Summary

This chapter focuses on addressing climate change proactively by employing Big Earth Data methodologies to calculate the greenhouse gas emissions indicator (SDG 13.2.2) and provide corresponding spatiotemporal data products. Based on the data and findings presented, we find that global greenhouse gas emissions are still under great pressure. In contrast, China's terrestrial ecosystem carbon sinks exhibit a more significant increasing trend compared to the global average, resulting in a notable rise in its contribution to global carbon sinks since 1980. Regarding indicator evaluation, we suggest modifying the SDG 13.2.2 indicator to better assess countries of varying scales by adopting metrics such as per capita greenhouse gas emissions or emissions per unit of GDP.

While both China and the global community have made progress in enhancing ecosystem carbon sinks under SDG 13, future efforts must prioritize monitoring greenhouse gas reduction implementations. On the one hand, it is essential to improve data monitoring capabilities for greenhouse gas emissions. On the other hand, strong policies must be formulated to accelerate emission reductions while simultaneously enhancing forest conservation and farmland management to increase ecosystems' capacity to sequester greenhouse gases.

References

United Nations (2015) Transforming our world: the 2030 Agenda for sustainable development [R]. https://sustainabledevelopment.un.org/post2015/transformingourworld [2024-10-09]

United Nations (2022) The sustainable development goals report 2022 [R]. https://desapublications.un.org/publications/sustainable-development-goals-report-2022 [2023-10-09]

WEF (2023) The global risks report 2023. 18th Edition [R]. https://www3.weforum.org/docs/WEF_Global_Risks_Report_2023.pdf [2023-10-01]
WMO (2023) State of the global climate 2022 [R]. https://wmo.int/publication-series/state-of-global-climate-2022 [2023-10-01]
Yuan WP, Liu D, Dong WJ et al (2014) Multiyear precipitation reduction strongly decreases carbon uptake over Northern China [J]. J Geophys Res Biogeosci 119(5):881–896

7 SDG 14 Life Below Water

7.1 Background

The ocean plays a crucial role in regulating the global water cycle, controlling the climate, protecting biodiversity, and providing habitats for many important species. In addition, marine products supply at least 20% of animal protein to 3.1 billion people worldwide, which is particularly important for the livelihoods of coastal regions and small island developing countries with poorer economic conditions.

The UN's 2030 Agenda includes SDG 14, which aims to conserve and sustainably use the oceans and marine resources, as part of its 17 transformative goals. However, from a global perspective, the implementation of many specific targets under SDG 14 has been less than satisfactory. In the report titled "Progress towards the Sustainable Development Goals: Towards a Rescue Plan for People and Planet: Report of the Secretary-General (Special Edition)" presented on April 27, 2023, the UN Secretary-General summarized the latest progress on SDG 14. The overall assessment shows that although some progress has been made globally in expanding marine protected areas and combating illegal, unreported, and unregulated (IUU) fishing, the trends that harm ocean health have not diminished. The oceans continue to face threats from acidification, eutrophication, declining fish populations, and increasing plastic pollution. About 50% of the indicators are either stagnating or regressing, highlighting the urgent need for more coordinated and accelerated efforts (United Nations 2023b).

This chapter focuses on a deep analysis of specific targets under SDG 14.1 and SDG 14.2, which have methods but lack data. By examining the distribution and migration patterns of microplastics in the Arctic-Pacific sector, the temporal and spatial changes of phytoplankton blooms in the North Indian Ocean, and the dynamic monitoring of coral cover on typical coral reef islands, this chapter explores methods for obtaining relevant data and computational models for typical global regions. This chapter aims to fill the data gaps in the UN's reports and provide a better understanding of the progress on SDG 14, thereby facilitating the implementation of measures to achieve these goals.

7.2 Main Contributions

This chapter conducts monitoring and evaluation of SDG 14.1 and SDG 14.2 related indicators in China and surrounding regions through three case studies. The main contributions are shown in Table 7.1.

H. Guo, *Big Earth Data in Support of the Sustainable Development Goals (2023)—The Belt and Road*, Sustainable Development Goals Series, https://doi.org/10.1007/978-981-95-3178-3_7

Table 7.1 Cases and their main contributions

Targets	Tiers	Cases	Contributions
SDG 14.1 By 2025, prevent and significantly reduce marine pollution of all kinds, in particular from land-based activities, including marine debris and nutrient pollution	Tier II	Distribution and Migration Characteristics of Microplastics in the Arctic-Pacific Sector	**Data product**: A basic dataset of microplastic characteristics in multiple environmental media in the Arctic-Pacific sector **Method and model**: Establish chemical and physical characteristic indicators of microplastics and express their occurrence patterns in the environment **Decision support**: Connecting the construction of the "Ice Silk Road" will help explore effective new mechanisms for Arctic environmental governance
SDG 14.2 By 2020, sustainably manage and protect marine and coastal ecosystems to avoid significant adverse impacts, including by strengthening their resilience, and take action for their restoration in order to achieve healthy and productive oceans	Tier II	Spatiotemporal Variations of Large-Scale Phytoplankton Blooms in the North Indian Ocean	**Data product**: Inverted atlas of large-scale phytoplankton algae blossoms ($\geq$ 4 km scale) in the North Indian Ocean from 2015 to 2022 **Method and model**: Remote sensing identification method for phytoplankton blooms in the North Indian Ocean **Decision support**: Providing algal bloom monitoring methods and data support for the North Indian Ocean and other sea areas, serving the monitoring and prevention of algal bloom disasters in the Belt and Road sea areas such as the North Indian Ocean, which is in line with the goal of the SDG 14.2 to strengthen disaster resistance, sustainable management and protection of marine ecosystems
		Dynamic Monitoring of Live Coral Cover on Typical Islands and Reefs	**Data product**: Using satellite imagery from the Landsat Collection2 Level-2 dataset and Sentinel-2_MSI_L2A dataset from 1998 to 2022 **Method and model**: On the basis of excluding areas with obvious inactive coral distribution such as deep water zones and underwater sandy areas, the relationship between remote sensing reflectance and active coral coverage is fitted using band combinations and quadratic polynomial models **Decision support**: Provide support for coral reef ecological management; provide information security for research on global climate change, sustainable development, etc

7.3 Case Study

7.3.1 Distribution and Migration Characteristics of Microplastics in the Arctic-Pacific Sector

Target: SDG 14.1 By 2025, prevent and significantly reduce marine pollution of all kinds, in particular from land-based activities, including marine debris and nutrient pollution.

- **Background**

Since the large-scale production of plastics, their global production has shown an exponential growth trend (Borrelle et al. 2020), and the total global plastic production in 2022 reached 400 million tons (Plastics Europe 2023). The improper handling and fragmentation of environmental plastic waste have led to a continuous increase in microplastics (MPs) pollution (microplastics with particle sizes less than

5 mm). More and more studies have shown that microplastics are widely present in the global marine environment (Lima et al. 2021) and pose significant risks to the ecological environment and human health. Therefore, marine microplastics, as a new type of pollutant, have received high attention from scholars and the public both domestically and internationally, and the level of attention continues to rise. The Arctic Ocean is an important "window" of global environmental change. It also occupies a special strategic position in the "Polar Silk Road". At the same time, there are famous fishing grounds and unique biological species. Researchers found that the branches of the Arctic hot salt circulation can bring plastic waste from other oceans to the marginal sea of the Arctic Ocean, and put forward the view that the seabed of the Arctic Ocean may be an important "sink" of plastic waste (Cózar et al. 2017; Obbard et al. 2014; Peeken et al. 2018). However, the environmental problems of microplastics in the Arctic Ocean are not limited to scientific problems, but also can be extended to one of the focuses of global governance, economy and politics. So far, we have little research on microplastics in the seawater of the Arctic-Pacific sector, and lack a comprehensive understanding of the pollution status and the fate of microplastics in the Arctic Ocean.

Within the framework of the SDGs of the 2030 Agenda, reducing marine debris is one of the important indicators for the prevention and control of "marine pollution". Therefore, this case study focuses on the UN SDGs 14.1 and uses published data and scientific investigation data to carry out research on the current situation and distribution characteristics of microplastic pollution in different media in the Arctic-Pacific sector, so as to provide basic data for the assessment of microplastic pollution, migration and environmental impact in the Arctic-Pacific sector. This case study will also help to link up the relevant international policies on environmental protection in the polar regions and support the research on the SDGs of the UN.

- **Data**
  - Literature data on microplastics in the Arctic-Pacific sector published in international journals. International journals mainly include: *Nature Communications*, *Environmental Research*, *Environment International*, *Chemosphere*, and *Marine Pollution Bulletin*.
  - Scientific investigation and monitoring data, such as the 8th and 9th scientific expeditions in the Arctic.

- **Method**

Based on the data published in international journals and scientific investigation data, the occurrence state of microplastics in water, sediments and typical organisms in the Arctic-Pacific sector was analyzed by using the abundance, particle size and polymer types of marine microplastics. The correlation of microplastics in different media was studied, and the spatial distribution pattern of microplastics in the Arctic-Pacific sector was systematically analyzed. Combined with the action of ocean currents, the spatial pattern of microplastics in the sea area and the influence of ocean current transport were explored. In this case study, microplastic samples in seawater were mainly collected by pump and trawl. Microplastics in sediments were extracted by density flotation. Microplastics in biology were separated after digestion by weak oxidation. The polymer types of microplastics were identified by using a micro-Fourier Transform Infrared Spectroscopy.

- **Results and Analysis**

1. **Distribution Characteristics of Microplastics in Seawater**

The types of microplastics in the seawater of the Arctic-Pacific sector mainly included rayon, polyethylene terephthalate (PET), polyacrylamide (PAM), chlorinated polyethylene (CPE), polyethylene (PE), polyester fiber, poly(vinyl alcohol) (PVA), polyvinyl chloride (PVC),

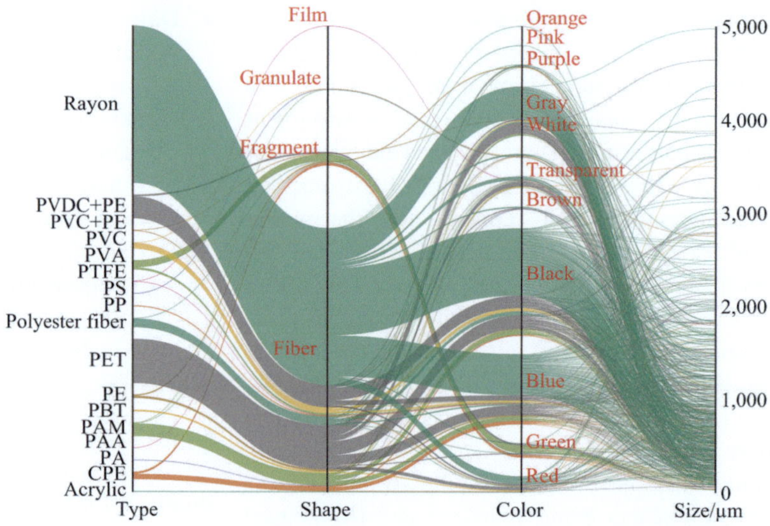

Fig. 7.1 Distribution characteristics of microplastics in the seawater of the Arctic-Pacific sector

polypropylene (PP), polystyrene (PS), polytetrafluoroethylene (PTFE), polyamide (PA), poly(acrylic acid) (PAA), polyvinylidene chloride (PVDC), polybutylene terephthalate (PBT) and acrylic, among which PA (56.21%) and PET (15.83%) accounted for a relatively high proportion. The proportions of all detected plastic particles with particle sizes of 0.0–1.0 mm and 1.0–2.0 mm were relatively high, which were 69.82% and 21.75% respectively. A total of 11 colors of microplastics were detected, of which black (38.46%), blue (25.15%) and gray (17.46%) accounted for a high proportion, followed by red (5.77%), green (5.18%) and transparent color (4.18%). The shape of microplastics was mainly fibrous, accounting for 94.53%, with minor fragment, granular and film (Fig. 7.1).

The abundance of microplastics in all study areas ranged from N.D.-260.57 items/m^3, with an average abundance of 34.31 items/m^3 and a median of 21.20 items/m^3. The average abundance of microplastics in the Arctic Ocean was 34.52 items/m^3, and the average abundance of microplastics in the Pacific sector was 22.25 items/m^3. On the whole, there were differences in the abundance distribution of microplastics between the Arctic Ocean and the Pacific sector. High microplastic abundance appeared in the Beaufort vortex area. It can be seen that the ocean current vortex convergence had an agglomeration-driving effect on microplastics (Ross et al. 2021) (Fig. 7.2).

2. **Distribution Characteristics of Microplastics in Sediments**

The abundance range of microplastics detected in the sediments of the study sea area was 5.30 items/kg-220.00 items/kg, and the average abundance was 83.24 items/kg. Seven types of microplastics were found, among which PET, PP and PS accounted for a relatively high proportion, accounting for 34%, 31% and 11% respectively, followed by rayon, PVC, PA and polyacrylonitrile (PAN). All detected microplastics were mainly fibrous (60%), fragment (22%) and film (18%). Microplastics with particle sizes ranging from 1.0 to 2.0 mm were relatively more, accounting for 47.0% of the total detected microplastics (Mu et al. 2019; Kanhai et al. 2019) (Fig. 7.3).

From the spatial distribution characteristics of microplastics, it was found that the abundance of microplastics in the sediments close to the central Arctic Ocean was high, indicating that there was a "sink" phenomenon of microplastic deposition in the study area. In addition, there were many low-density microplastics in the

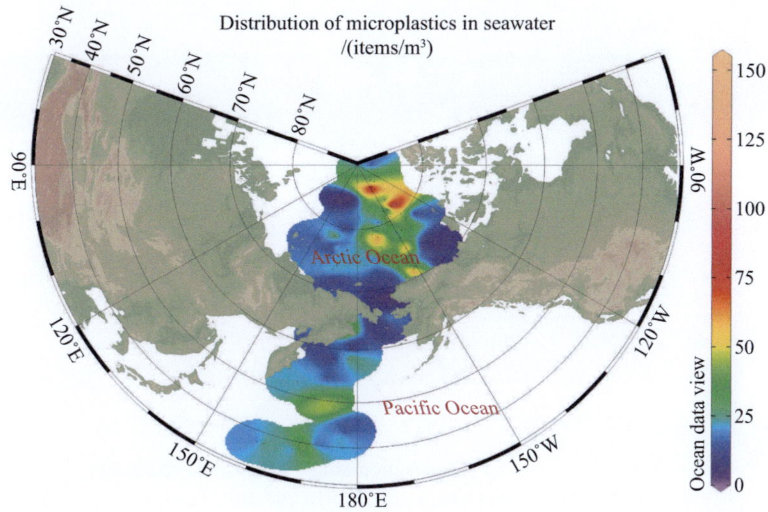

Fig. 7.2 Spatial distribution of microplastics in the seawater of the Arctic-Pacific sector

sediments, which revealed the phenomenon of the vertical transport mechanism of water to microplastics in the area, indicating that it had a certain driving effect on the vertical migration of microplastics.

3. **Distribution of Microplastics in Benthos**

Six typical benthos in the Arctic-Pacific sector were analyzed (Fang et al. 2018, 2021), including whelks (*L. hypolispus, R. daphnelloides* and *E. nana*), shrimp (*P. borealis*), crab (*C. opilio*), starfish (*A. rubens, C. crispatus* and *L. polaris*), brittle star (*O. sarsii*) and bivalves (*A. crenata* and *M. tokyoensis*). The abundance of microplastics in all benthos ranged from 0.18 items/ind. to 0.83 items/ind., and the average abundance of microplastics was 0.58 items/ind. The main types of microplastics in organisms were PA, PE, PET and cellophane (CP), with percentages of 46.80%, 22.41%, 24.25% and 6.53% respectively. There were many microplastics with particle sizes ranging from 0.1 to 1.5 mm, accounting for about 69.53% (Fig. 7.4).

It was found that there were some differences in the enrichment of microplastics by different individuals of benthos. The enrichment characteristics of microplastics in organisms of different species were not only affected by the distribution of microplastics in their habitat, but also related to their living habits and feeding patterns. The spatial distribution of high concentration microplastics in sediments and organisms in the study area was consistent, which were distributed in the ocean near the center of the Arctic Ocean. Combined with other studies, it was preliminarily inferred that Arctic sea ice and ocean currents had a certain impact on the migration and agglomeration of microplastics.

4. **Relationship Between the Spatial Distribution of Microplastics and Ocean Currents**

The surface current in the Arctic Ocean is mainly composed of Transpolar Drift and clockwise Beaufort Gyre, which control the direction of ice flow and modern sedimentary model in the Arctic Ocean. The Arctic Ocean communicates with the Atlantic Ocean and the Pacific Ocean through ocean currents. The seawater of the Pacific Ocean enters the Arctic Ocean through the Bering Strait, and the seawater flow is about 1 Sv. The Atlantic Ocean current is divided into two branches, which enter the Arctic Ocean through the Barents Sea and the Fram Strait, with a flow of about 3 Sv and 6.6 Sv respectively. Seawater from the Pacific Ocean and the Atlantic Ocean eventually flows out of the Arctic Ocean through the Davis Strait and Fram Strait (Rudels 2001; Polyakov et al. 2012) (Fig. 7.5).

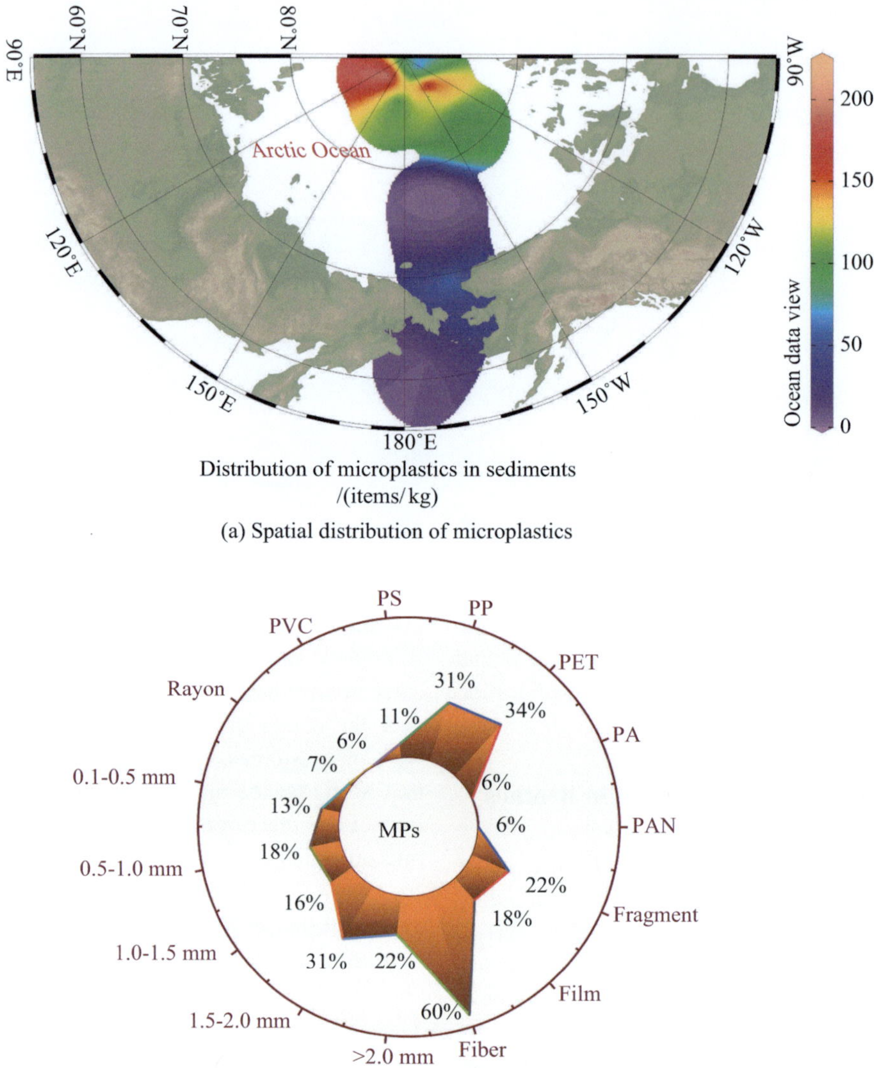

(a) Spatial distribution of microplastics

(b) Physical and chemical indicators of microplastics

Fig. 7.3 Spatial distribution and index characteristics of microplastics in sediments

The spatial distribution characteristics of microplastics in the Arctic Ocean showed that the high concentration of microplastics in the seawater appeared in the Beaufort vortex area and in the intersection area between the Transpolar Drift and the West Spitsbergen Current from the Atlantic (the area near the Svalbard Islands). The results further proved that the vortex and ocean current accumulation area in the Arctic Ocean could lead to the agglomeration of microplastics, showing a high abundance of microplastics.

Note: The green arrow indicates warm North Atlantic water. The purple arrow is cold, low-salt polar water. The red arrow is low-salt denatured water. AC: Anadyr Current, ACC: Alaskan Coastal Current, BC: Baffin Current, BIC: Bear Island Current, BG: Beaufort Gyre, EGC: East Greenland Current, EIC: East Iceland Current, ESC: East Spitsbergen Current, IC: Irminger Current, JMC: Jan Mayen Current, MC: Murman Current, NAD: North Atlantic Drift, NAC: Norwegian Atlantic Current, NCC:

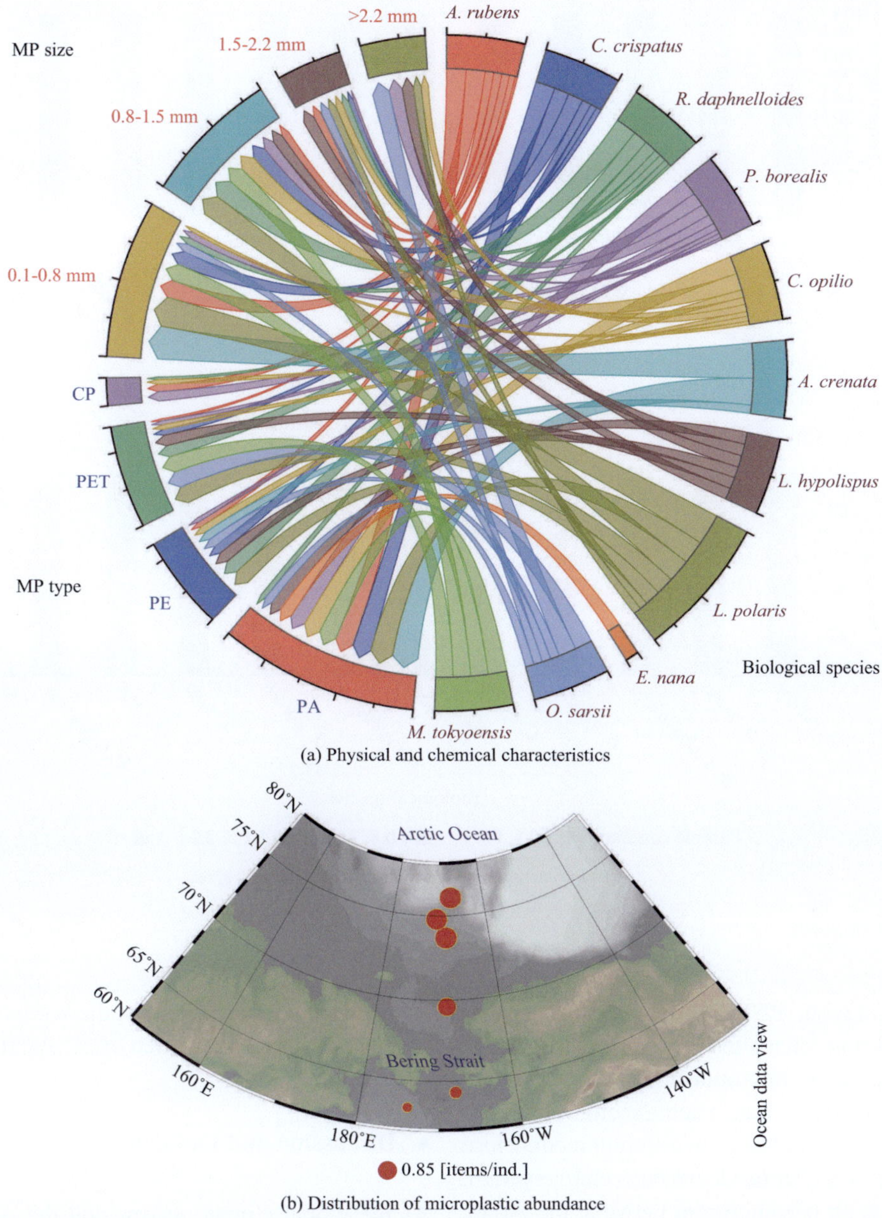

Fig. 7.4 Characteristics and distribution of microplastics in benthos

Norwegian Coastal Current, SB: Siberian branch (of the Transpolar Drift), SCC: Siberian Coastal Current, TPD: Transpolar Drift, WGC: West Greenland Current, WSC: West Spitsbergen Current.

Highlights

- The types of microplastics in seawater, sediments and organisms in the Arctic-Pacific sector were diversified. Microplastics were

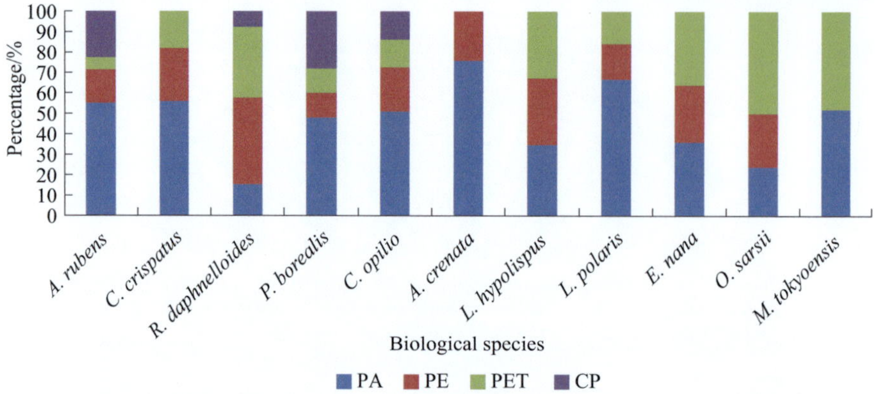

(c) Percentage contents of microplastics

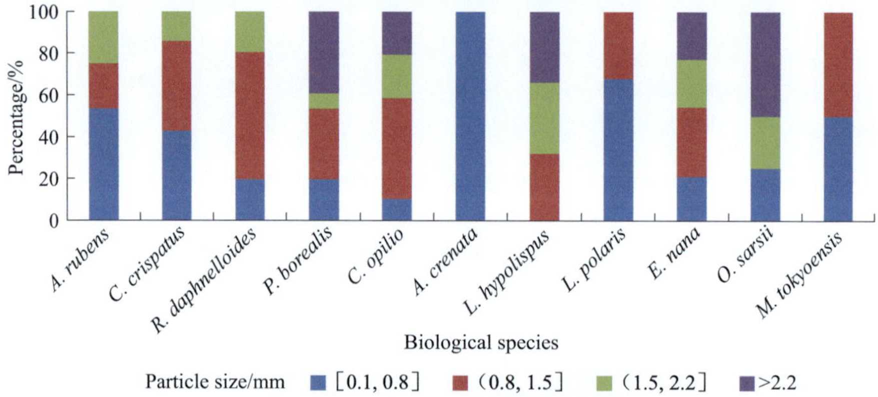

(d) Percentage of MPs with different particle sizes

Fig. 7.4 (continued)

mainly rayon, PET, PP and PS. The spatial distribution characteristics of microplastics showed that microplastic abundance was different in the Arctic-Pacific sector. The abundance of microplastics in different media close to the central Arctic Ocean was relatively high.

- There is an ocean current between the Arctic Ocean, the Atlantic Ocean and the Pacific Ocean, and the seawater runs through and exchanges with each other. The surface current in the Arctic Ocean is mainly composed of the Transpolar Drift and the clockwise Beaufort Gyre. The high concentration of microplastics in the Arctic Ocean water appeared in the Beaufort vortex area and the intersection area between the Transpolar Drift and the West Spitsbergen Current from the Atlantic, which further proved that the agglomeration of microplastics can occur in the vortex area and current convergence area of the Arctic Ocean.

- **Discussion and Outlook**

As the global climate warms and the Arctic sea ice melts, the Arctic ecological environment is an important issue of concern to all stakeholders. Arctic environmental issues, including Arctic climate, marine life and environmental protection, are still in the initial stage of cognition, exploration and discussion. The Arctic Ocean and the Pacific sector are closely connected as a whole through the circulation system. Because the density of the incoming seawater from the Pacific Ocean is less than that from the Atlantic Ocean, the seawater from the Pacific sector

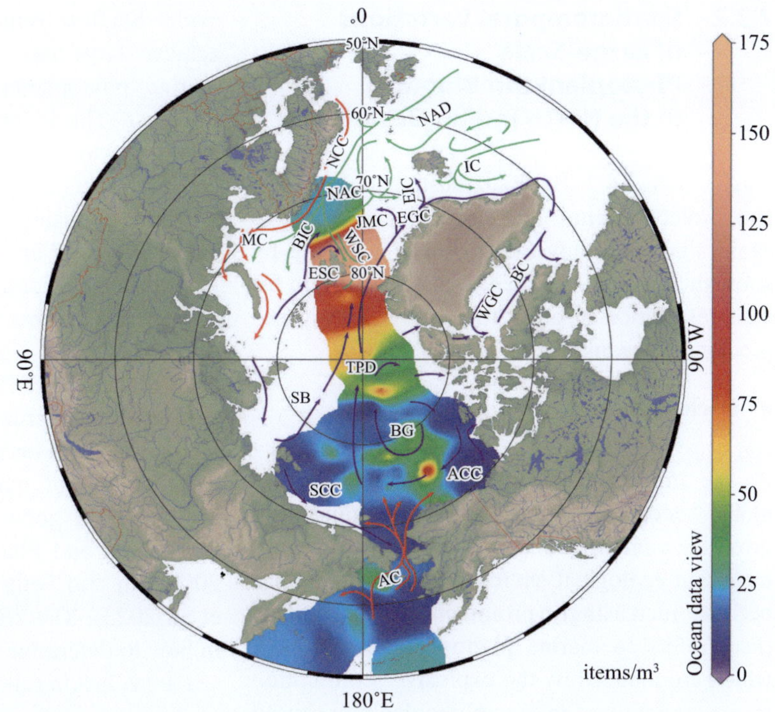

Fig. 7.5 Relationship between microplastic distribution in water bodies and ocean currents

will mainly enter the upper layer of the Arctic Ocean, which will have a more direct impact on the climate and environment of the Arctic Ocean. The environmental problems of the Arctic Ocean are related to the common interests of mankind. The environmental protection of the Arctic Ocean has the characteristics of borderlessness and globalization. The governance and protection of the Arctic Ocean environment require global solidarity and cooperation. Therefore, knowing and understanding the environmental status and development trend of the Arctic Ocean is the premise of effectively participating in Arctic environmental governance and safeguarding environmental rights and interests.

In order to promote and apply the spirit of the "Silk Road" in the Arctic region and the Arctic Ocean, the construction of the "Ice Silk Road" is the deepening and extension of the connotation of the BRI. From the perspective of global environmental protection and the maintenance of the common interests of mankind, it is of great significance to prevent, reduce and control the pollution of the Arctic marine environment. The research showed that the source, distribution and harm of marine microplastics had global significance. The irreversibility of environmental pollution, ecological risk and health risk is uncertain. Therefore, studying the occurrence form and spatial distribution characteristics of microplastics in the Arctic environment, and exploring the spatial migration process and migration law of microplastics, can predict the convergence and fate of marine microplastics in the environment, which could provide theoretical guidance for the treatment of marine microplastic pollution. In addition, in the future, we need to join hands with near-Arctic countries to build an environmental community and jointly explore effective new mechanisms for Arctic environmental governance.

7.3.2 Spatiotemporal Variations of Large-Scale Phytoplankton Blooms in the North Indian Ocean

Target: SDG 14.2: By 2020, sustainably manage and protect marine and coastal ecosystems to avoid significant adverse impacts, including by strengthening their resilience, and take action for their restoration in order to achieve healthy and productive oceans.

- **Background**

As the importance of the marine economy in global economic development continues to grow, the contradiction between marine development and ecological environment protection has become increasingly prominent. Algal blooms (here refers to marine phytoplankton blooms), which are induced by the explosive proliferation and aggregation of marine phytoplankton caused by water eutrophication, are one of the most common ecological disasters when the environment is overloaded. Marine phytoplankton blooms resulting from the outbreak of toxic algae will release toxins, leading to the death of marine organisms and further exacerbating the deterioration of the marine environment. While the excessive reproduction of non-toxic algae will cause a decrease in the oxygen content in the ocean, resulting in asphyxiation of marine organisms such as fish due to a lack of oxygen, which also triggers the deterioration of the marine environment. Therefore, the outbreak of phytoplankton blooms will bring serious harm to the health of coastal ecosystems, fishery development, the safety of cooling sources for coastal nuclear power plants, and even human health and social development. Under the combined influence of human activities and climate change, the frequency and scale of nearshore marine phytoplankton blooms have shown a trend of increasing year by year (Dai et al. 2023).

Compared with traditional on-site monitoring methods, satellite remote sensing has the advantages of wide coverage and short revisit cycle. Satellite remote sensing monitoring can achieve large-area synchronous observation of marine phytoplankton blooms at a relatively low cost. The identification methods of marine phytoplankton blooms based on the abnormal concentration of chlorophyll a (Chla) retrieved by remote sensing have been widely applied in ocean waters (Wang et al. 2021; Dai et al. 2023). In addition, spectral index algorithms such as normalized fluorescence line height (nFLH), red tide index (RI), algal bloom index (RBI), red band difference (RBD), and the newly proposed commission internationale del'éclairage (CIE) method have all been successfully applied to the identification of marine phytoplankton blooms and achieved good retrieval results (Hu et al. 2005; Ahn and Shanmugam 2006; Amin et al. 2009; Hu and Feng 2017; Liu et al. 2022; Dai et al. 2023). The key to the above methods lies in how to determine the thresholds. Especially in sea areas lacking measured data, the application of these methods is subject to certain limitations. Recently, some studies have applied the 90th percentile threshold method, which is used for identifying heatwaves (i.e., extreme high-temperature events) in physical oceanography, to the identification of strong (high-concentration) phytoplankton bloom events. It has been applied to the identification of the climatic change of remote sensing chlorophyll a concentration data to identify strong (high-concentration) phytoplankton bloom events in the South China Sea (Lu et al. 2022). Herein, this case study proposes a novel approach integrating the 90th percentile threshold method with the typical colorimetry method, where intense phytoplankton bloom signals identified by the former serve as the basis for calculating thresholds in the latter for large-scale phytoplankton bloom detection. This strategy enables automated threshold determination in regions lacking prior bloom knowledge, thereby paving the way for automated dynamic bloom monitoring.

The North Indian Ocean, a critical sea area under the BRI, hosts extensive international collaboration among multiple partner nations. It is affected by dust storms and tropical cyclones all year round, thus facing great pressure on the marine environment and frequently suffering from phytoplankton bloom disasters. Clarifying the long-term dynamic changes of marine phytoplankton blooms in the North Indian Ocean is an important guarantee for the ecologically sustainable development and construction of the "Belt and Road". It is also a key link in the dynamic monitoring, prevention and control of marine ecological disasters (algal blooms) as well as the health evaluation of marine ecosystems, which meets the indicator requirements of SDG 14.2. Currently, there are relatively few studies on the identification and long-term analysis of phytoplankton blooms in the North Indian Ocean. The spatiotemporal distribution characteristics of phytoplankton blooms in the North Indian Ocean remain to be revealed, and there is also a lack of the release of relevant achievement progress. Therefore, from the perspective of promoting the SDGs through scientific and technological innovation, the research on the dynamic distribution and causes of large-scale phytoplankton blooms in the North Indian Ocean has scientific value in aspects such as the formation mechanism of algal blooms and withstanding the impact of meteorological disasters on the marine ecological environment. It conforms to the goal of strengthening disaster resilience in SDG 14.2 and is of great significance for the achievement of the marine SDGs.

- **Data**
  - The Level 3 (L3) Chla concentration data, photosynthetic active radiation (PAR) data, and sea surface temperature (SST) data were integrated from MODIS Aqua and Terra from 2003 to 2022, with a temporal resolution of 1 day and a spatial resolution of 4 km.
  - The model output nitrate (NO_3) data from 2015 to 2020, with a time resolution of 1 day and a spatial resolution of 0.25° × 0.25°, were downloaded from the Copernicus Marine Environment Monitoring Service (CMEMS).
  - The daily WindSat wind field data at 10 m above the sea level from 2015 to 2022, with a spatial resolution of 0.25° × 0.25°, were obtained from Remote Sensing Systems.

- **Method**

1. Phytoplankton Bloom Retrieval Method Based on the Colorimetry Method and the 90th Percentile Threshold Method

Due to the difficulty in obtaining measured data of phytoplankton blooms in the North Indian Ocean, the 90th percentile threshold method (Lu et al. 2022) was adopted to extract phytoplankton bloom data, which was used as a representation of the occurrence of strong phytoplankton bloom events. These data were then used to determine the angular thresholds required by the colorimetry method for retrieving phytoplankton blooms (Liu et al. 2022; Dai et al. 2023), in order to achieve the joint retrieval of phytoplankton blooms in the North Indian Ocean. Since the spatial resolution of the products used in this case study is 4 km and the determination of the chromatic angle thresholds is based on the information of strong phytoplankton blooms, the phytoplankton blooms identified in this case study are regarded as large-scale (large-area and high-concentration) phytoplankton blooms.

The specific calculation of the 90th percentile threshold method for retrieving strong phytoplankton blooms is as follows: Collect the Chla concentration data of all years in adjacent 11-day periods for each grid in the MODIS satellite data (including the data from 2003 to 2022), and form sample sets for each grid from these data grid by grid; Calculate the values located at the 90th percentile of the probability distribution in each sample set, and use them as a threshold for the occurrence of phytoplankton blooms in each grid; Subsequently, use a 31-day moving window to smooth the threshold sequence of each grid for each year; Finally, determine whether the Chla concentration at each grid point is continuously higher than the smoothed threshold for more than 3 days. If so, it is determined that a phytoplankton bloom event occurred at that grid (Lu et al. 2022).

$$R_{90\%} = \text{smooth}(90\%(\text{Chla}_{11\text{ days of all years}}), 31)$$

$$R_{90\%} \geq 3\text{days}$$

where, $\text{Chla}_{11\text{ days of all years}}$ represents the samples composed of the Chla concentration data in an adjacent 11-day window of each day for all years.

The specific steps of the CIE methods for phytoplankton bloom identification are as follows:

$$x = \frac{X}{X+Y+Z} - \frac{1}{3}$$

$$y = \frac{Y}{X+Y+Z} - \frac{1}{3}$$

$$X = 2.7689R + 1.7517G + 1.1302B$$

$$Y = 1.0000R + 4.5907G + 0.0601B$$

$$Z = 0.0000R + 0.0565G + 5.5943B$$

$$\alpha = \text{atan}\left(\frac{y}{x}\right)$$

where, *R*, *G*, and *B* respectively represent the red band (667 nm), green band (531 nm), and blue band (443 nm) of MODIS remote sensing data; *x* and *y* respectively denote the coordinates after the conversion of the *R*, *G*, and *B* color coordinate systems into the chromaticity angle coordinate system; α represents the chromaticity angle threshold judged according to the 90th percentile threshold method ($\alpha = 100°$).

2. **Grid-By-Grid Multiple Regression Statistical Analysis**

The spatial resolution of each element was resampled to be consistent with the spatial resolution (4 × 4 km) of the retrieval results of large-scale phytoplankton blooms. For each element, the climatological monthly mean value of each corresponding grid was subtracted to remove the seasonal signals. After standardizing the data with the seasonal signals removed, a multiple linear regression between each element and the retrieval results of large-scale phytoplankton blooms was established, and a qualitative analysis of the impact magnitude of each element on the algal blooms was conducted.

- **Results and Analysis**

Using the L3 Chla concentration data integrated from MODIS Terra and Aqua, the spatiotemporal distribution maps of large-scale phytoplankton blooms in the North Indian Ocean from 2015 to 2022 were jointly extracted based on the colorimetry method and the 90th percentile threshold method (Fig. 7.6). The northern part of the Arabian Sea is the most prominent area of large-scale phytoplankton blooms in the North Indian Ocean, where large-scale phytoplankton blooms exist all year round. Significant large-area phytoplankton blooms also exist in the Arabian Sea basin. Narrow strip-shaped phytoplankton blooms successively appear in the sea areas such as the east coast of Somalia, the Gulf of Mannar, and the east coast of the Bay of Bengal. Large-scale phytoplankton blooms also exist in the Arabian Sea basin. By multiplying the number of phytoplankton bloom pixels by the corresponding resolution (4 × 4 km), it is estimated that the annual area of large-scale phytoplankton blooms in the North Indian Ocean is approximately 4.7×10^{12} m^2. Meanwhile, the annual variations between the area and the intensity (i.e., Chla concentration) of these blooms are not completely synchronized. The intensity of large-scale phytoplankton blooms in the North Indian Ocean presents a gradually increasing trend, while the area of these blooms presents a weak decreasing trend (Fig. 7.6).

There is a high similarity in the annual variation trends between the number of days and the intensity of large-scale phytoplankton blooms in the North Indian Ocean (Fig. 7.7). Specifically, along the northern and eastern coasts of the

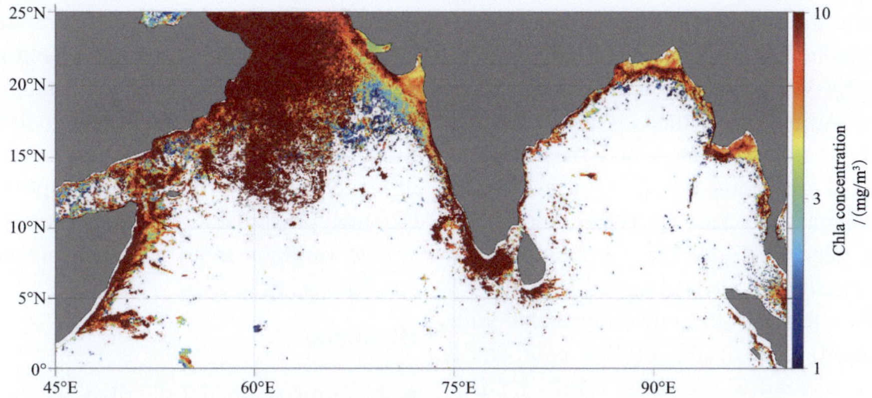

(a) Annual spatial distribution of Chla concentration for large-scale phytoplankton blooms

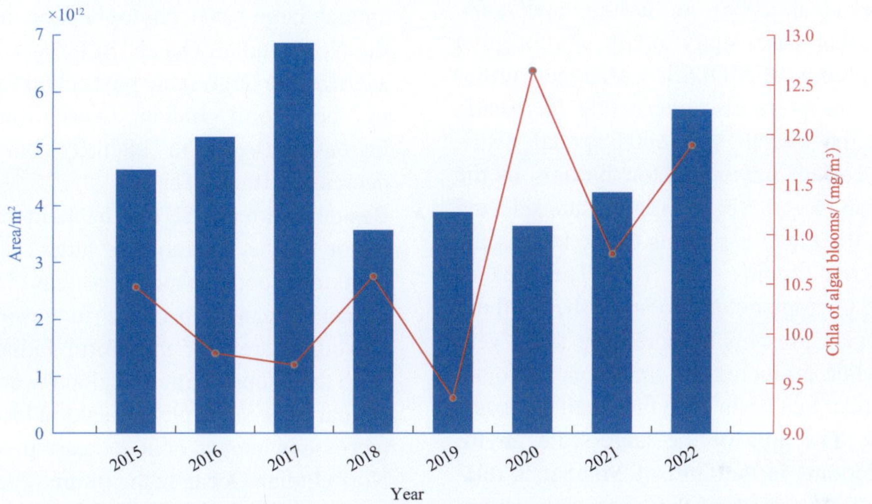

(b) Annual time series of the averaged area and Chla concentration for large-scale phytoplankton blooms

Fig. 7.6 Spatiotemporal distribution of large-scale phytoplankton blooms in the North Indian Ocean from 2015 to 2022

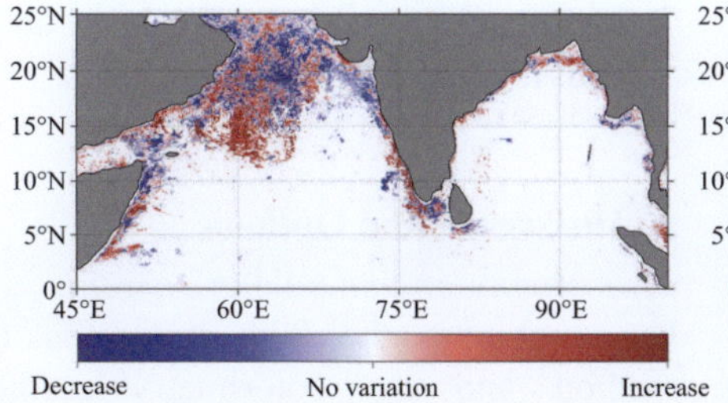

(a) Rate of bloom average intense from 2015-2022

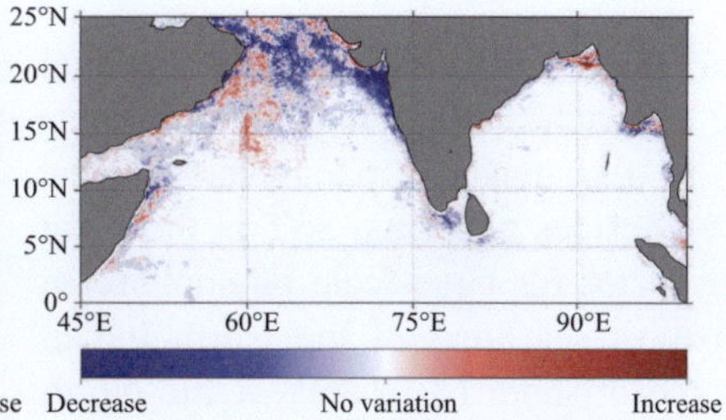

(b) Rate of bloom days from 2015-2022

Fig. 7.7 Trends of annual average intensity and days of phytoplankton blooms in the North Indian Ocean from 2015 to 2022

Arabian Sea, the number of days and the intensity of phytoplankton blooms show a decreasing trend year by year, while in the offshore basin area of the Arabian Sea, there is an increasing trend in the number of days and the intensity of phytoplankton blooms year by year. In the upwelling areas affected by river runoff, the increasing trends in both the number of days and the intensity of phytoplankton blooms are stronger than those in the upwelling areas with weaker runoff. The results of this case study are similar in trend to the latest results by Dai et al. (2023) along the coast of the North Indian Ocean (Dai et al. 2023). As a supplement, the results in this case study retain the original spatial resolution of MODIS, which can further reveal that there are also alternating increasing signals in the overall weakened spatial distribution of coastal phytoplankton blooms in the North Indian Ocean. Six key regions are selected to analyze their area variations of the large-scale phytoplankton blooms (Fig. 7.8). The area of large-scale phytoplankton bloom in the northern Gulf of Oman presents a decreasing trend year by year, while an increasing trend can be found in the Arabian Sea basin with the smallest occurrence area. The area of the large-scale phytoplankton blooms in the Gulf of Mannar is relatively small. The areas of the large-scale phytoplankton blooms on the east coast of Somalia, and on the north and east coasts of the Bay of Bengal are comparable, lying between those of the Gulf of Oman and the Gulf of Mannar (Fig. 7.8).

In the six key regions, nutrients play a crucial role in the occurrence of phytoplankton blooms compared to the dynamic parameters of the sea surface. Besides nutrients, SST is the main influencing factor for large-scale phytoplankton blooms in most sea areas of the North Indian Ocean. In the offshore basin area of the North Indian Ocean, the contribution of the wind field to phytoplankton blooms gradually exceeds that of SST, which is, to some extent, related to the fact that the region is affected by strong winds from dust storms and the nutrients brought by dust deposition induce phytoplankton blooms (Fig. 7.9). In several major upwelling areas in the North Indian Ocean, temperature plays a more important role in phytoplankton blooms compared to PAR and the wind field. With the intensification of global warming, in the North Indian Ocean, where temperature is an important influencing factor for large-scale phytoplankton blooms, phytoplankton blooms may show an annual variation trend of increasing in intensity but decreasing in area.

Highlights

- In terms of spatial distribution, the Northern Arabian Sea is the most prominent region with annual large-scale phytoplankton blooms in the North Indian Ocean. In terms of regional average, the large-scale phytoplankton blooms in the North Indian Ocean present an increasing trend in intensity but a weak decreasing trend in area.
- Besides nutrients, SST is the main influencing factor for the variations of large-scale phytoplankton blooms in most sea areas of the North Indian Ocean, especially in several major upwelling areas of the North Indian Ocean. With the intensification of global warming, the annual large-scale phytoplankton blooms may increase in intensity but decrease in area in the North Indian Ocean in the future. The research results provide scientific support for the formation of sea surface blooms in the North Indian Ocean and the governance of their ecological impacts, serving SDG 14.2 for strengthening disaster resilience and the sustainable management and protection of marine and coastal ecosystems.

- **Discussion and Outlook**

By combining the 90th percentile threshold method and the colorimetry method, this case study identified the large-scale phytoplankton blooms in the North Indian Ocean from 2015 to 2022 using the ocean color satellite remote sensing datasets, wind field datasets and nutrient reanalysis datasets. The spatiotemporal distribution characteristics of the large-scale phytoplankton blooms in the North

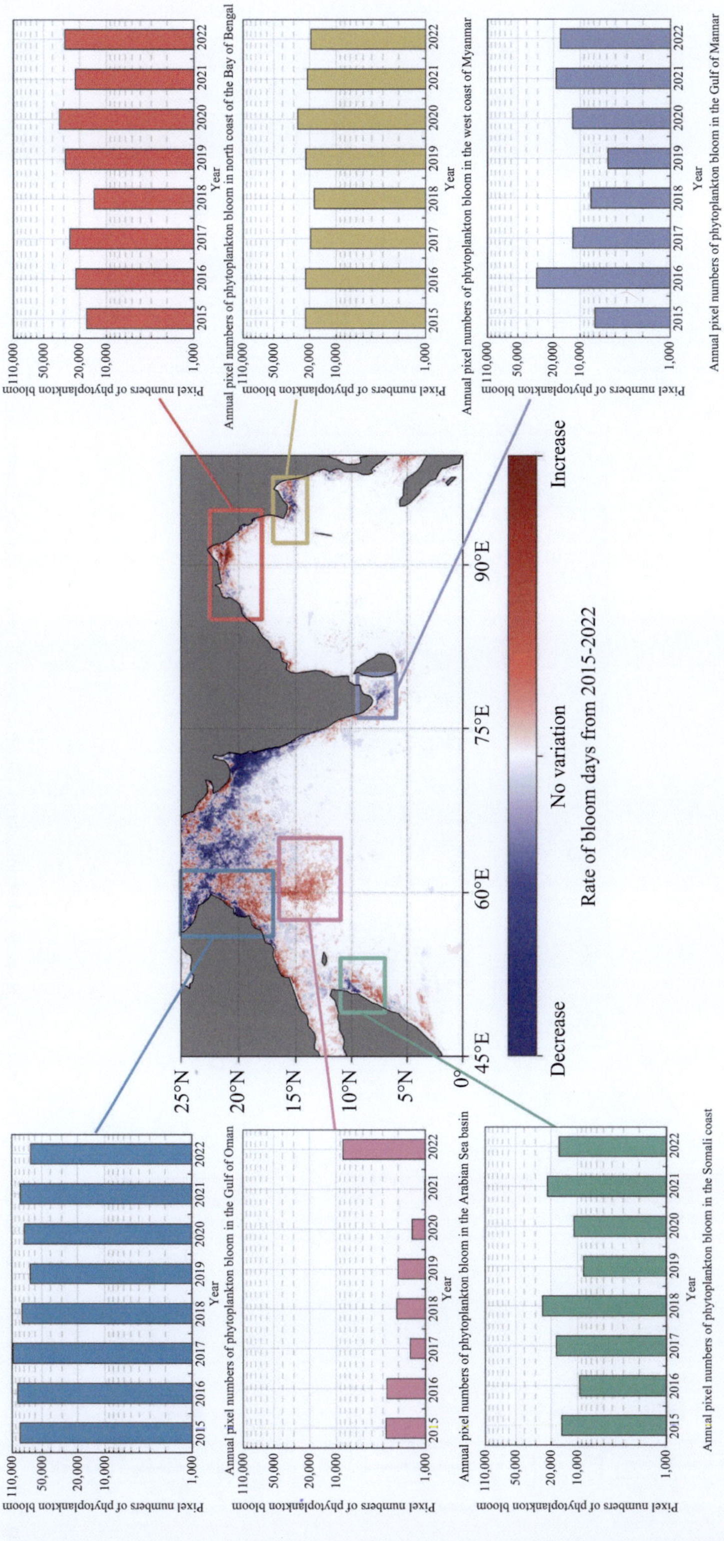

Fig. 7.8 Area variations of large-scale phytoplankton blooms in six key regions in the North Indian Ocean from 2015 to 2022

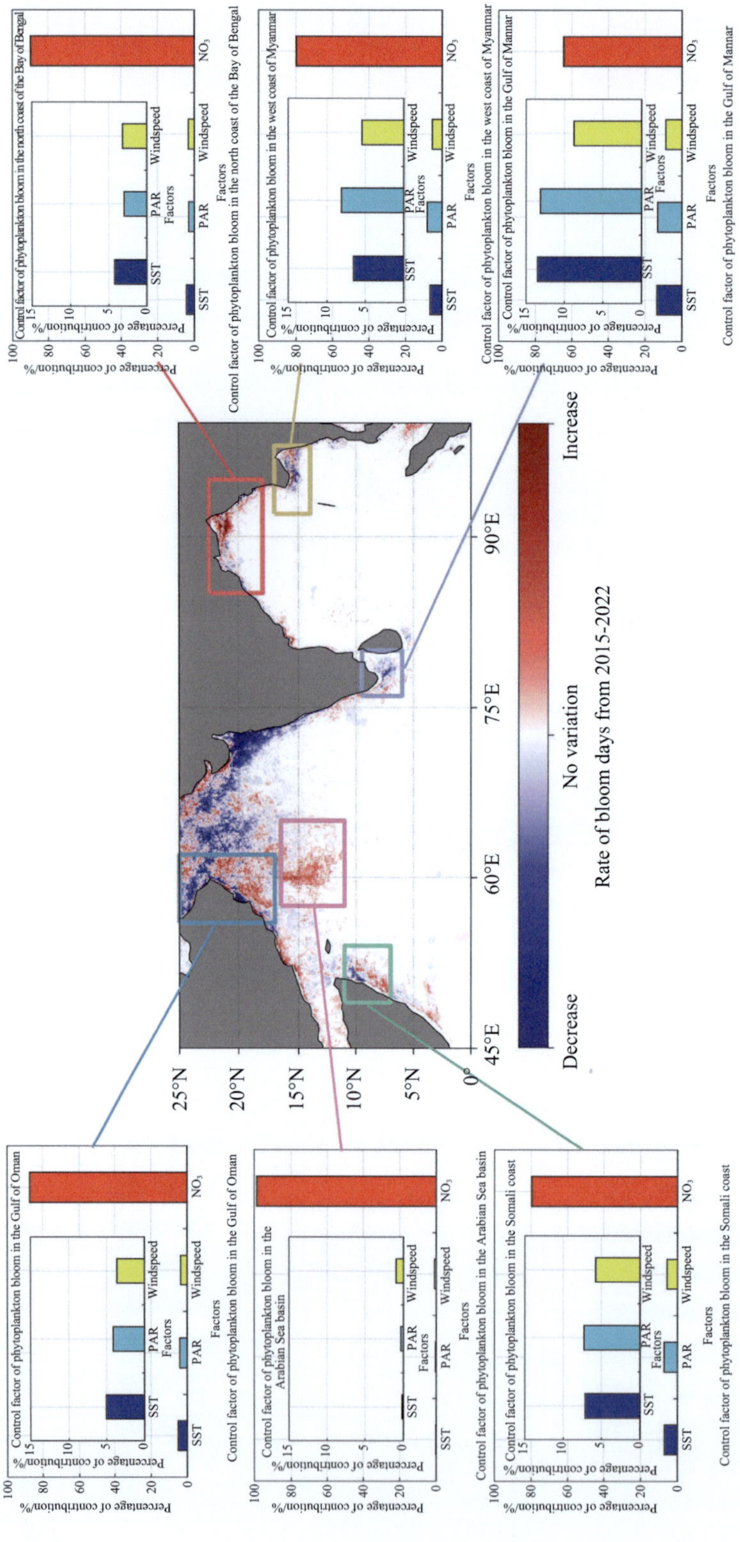

Fig. 7.9 Distribution of the main influencing factors for large-scale phytoplankton blooms at each grid in the six key regions in the North Indian Ocean from 2015 to 2022

Indian Ocean were revealed, and the possible causes of these blooms were explored. The results show that the Northern Arabian Sea is the main region for large-scale phytoplankton blooms in the North Indian Ocean. In addition, a significant annual increasing trend of both the number of days and intensity of large-scale phytoplankton blooms were found in the Arabian Sea basin, while a significant annual decreasing trend occurred along the eastern coast of the Arabian Sea. SST is the main influencing factor for most phytoplankton blooms in the North Indian Ocean besides nutrients, especially in several major upwelling areas. Therefore, with the intensification of global warming, large-scale phytoplankton blooms in the North Indian Ocean may show a trend of increasing in intensity but decreasing in area. The methods adopted in this case study can provide important technical support for aspects including algal bloom monitoring, prevention, and control, as well as the sustainable development of fisheries. It can also provide a decision-making basis and scientific services for the countries participating in the BRI to enhance the sustainable management and protection capabilities of coastal and marine ecosystems. Moreover, these results can provide important information support along the BRI for achieving the marine SDGs that are in line with SDG 14.2, which aims to improve marine management and protection capabilities.

To meet international demands, relevant research can incorporate more ecological and physical parameters in the future, and build an effective physical-ecological coupling model as well as a systematic ecological impact assessment system for phytoplankton blooms. In addition, in combination with artificial intelligence, it is possible to further achieve refined monitoring and prediction of algal blooms in the countries participating in the BRI. This will enhance the capabilities of institutions such as the Group on Earth Observations (GEO), the State Oceanic Administration (SOA), and the International Centre for Ocean Development (ICOD) in terms of algal bloom monitoring, prevention, and control. These advances will provide data support and decision-making services for improving the sustainable management and protection capabilities of marine and coastal ecosystems and reducing marine pollution, serving SDG 14.2.

7.3.3 Dynamic Monitoring of Live Coral Cover on Typical Islands and Reefs

Target: SDG 14.2: By 2020, sustainably manage and protect marine and coastal ecosystems to avoid significant adverse impacts, including by strengthening their resilience, and take action for their restoration in order to achieve healthy and productive oceans.

- **Background**

Against the backdrop of the rapid degradation of coral reefs worldwide, the coral reefs in the Xisha Islands have not been spared either, with their rate of degradation even exceeding the global average. For instance, the average coral coverage on Yongxing Island of the Xisha Islands dropped from 90% in 1980 to 20% between 2008 and 2009. In recent years, with the rapid development of China's economy, society, and national defense, the demand for space in the South China Sea's coral reefs has been increasing. Affected by global climate change, environmental pollution, overfishing, and unreasonable development activities, the degradation of coral reef ecosystems has become increasingly severe. This not only leads to a decline in biodiversity but also causes frequent natural ecological disasters, making it difficult to maintain the habitats of some islands and reefs.

However, current investigations of live coral cover (LCC) mainly rely on standard transect or photoquadrat methods, which are only suitable for surveys at typical sections and points. Due to the limited number of samples collected in actual surveys and the significant impact of the selection of transect and quadrat locations on the results, the uncertainty of investigation and monitoring outcomes has increased, making it difficult to adapt to comprehensive surveys of coral reefs. Therefore, it is all the more difficult to conduct regular on-site investigations of LCC, especially for large-scale coral reefs. As a result, countries urgently need to develop advanced coral reef ecological remote sensing monitoring methods to address the need for monitoring the ecological conditions of typical coral reef ecosystems.

- **Data**
  - Satellite images of the LandSat Collection 2 Level-2 dataset and Sentinel-2_MSI_L2A dataset from 1998 to 2022, downloaded from https://glovis.usgs.gov/app and https://dataspace.copernicus.eu, with LandSat images with a spatial resolution of 30 m, and Sentinel-2 images with a spatial resolution of 10 m.
  - Remote sensing images and published data from 2015 to 2022 Landsat operational land imager (OLI) and other multi-spectral satellites.

- **Method**

With the increasing abundance of medium- and high-resolution multi-spectral satellite resources, the use of these satellite remote sensing technologies for dynamic monitoring of LCC has shown great potential and wide application prospects (Liao et al. 2021). In this case study, on the basis of systematic analysis of spectral reflectance characteristics of typical corals (Zhao et al. 2013), the second-order difference operator of blue wave segments was considered, and the regression analysis of field ecological survey data and band ratio (Souter et al. 2021) was used to construct an LCC model. By analyzing the space scalability, time scalability and sensor scalability of the model, the accuracy of the LCC inversion model that we have established is relatively high, and its root-mean-square error and average absolute value error are 8.84% and 5.79%, respectively. This method can effectively extract LCC information and provide accurate reference for the health analysis and protection management of the coral reef ecosystem.

Using bands 1, 2 and 3 of Landsat5 images, bands 1, 2, 3 and 4 of Sentinel-2 and Landsat 8 images, LCC was estimated according to the following formula:

$$I = 0.75B_1 - 1.25B_2 + 0.25B_3 + 0.25B_4$$

$$\text{LCC} = 0.0001 \times I^2 + 0.0742 \times I + 14.4582$$

When the image is Landsat8, B_1, B_2, B_3 and B_4 are the first, second, third and fourth bands of the image, respectively. When the image is Landsat5, B_2, B_3 and B_4 are the first, second and third bands of the image respectively, $B_1 = B_2 + 200$, and LCC is the live coral cover.

- **Results and Analysis**

The LCC of typical islands and reefs is in a decreasing state. Using satellite data such as Landsat OLI and China's "Haiyang 1C/D" (HY-1C/D), the LCC of the Hawaii Reef (Kure Reef) (Fig. 7.10) and Great Barrier Reef (Nameless Reef) (Fig. 7.11) in 2015, 2020 and 2022 was calculated. Remote sensing monitoring results show that the reefs and coral reefs in these two typical regions generally have a trend of degradation, and from 2015 to 2020, the LCC showed a significant decline, which was clearly reflected in the fact that the live coral distribution area was shrinking, with the Hawaii Reef (Kure Reef) decreasing from 24.00% to 18.63%, a decrease of 5.37%, which is basically consistent with the results of the field ecological survey in the same period, a decrease of 6.00%. The Great Barrier Reef (Nameless Reef) decreased from 18.16% to 10.85%, a decrease of 7.31%, which is basically consistent with the results of the field ecological survey in the same period, a decrease of 6.80%. From 2020 to 2022, the Great Barrier Reef (Nameless Reef) showed weak signs of recovery, and the Hawaii Reef (Kure Reef) was in a decline, but the reduction slowed down.

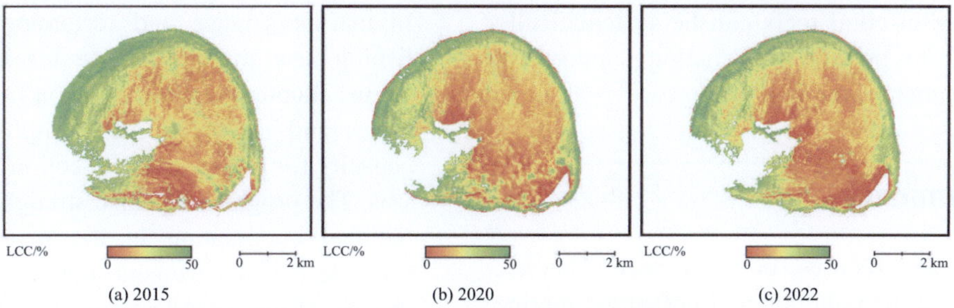

Fig. 7.10 LCC in the Hawaii Reef (Kure Reef) in 2015, 2020, and 2022

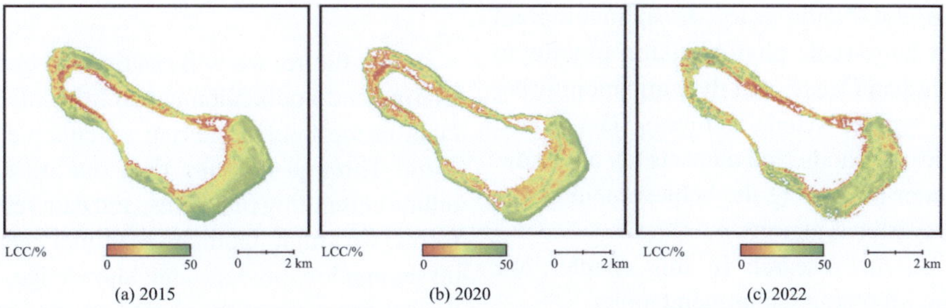

Fig. 7.11 LCC in the Great Barrier Reef (Nameless Reef) in 2015, 2020 and 2022

Highlights

- Use of multi-spectral satellite remote sensing images to invert LCC is of good accuracy. The model exhibits excellent scalability in terms of time, space, and sensors, effectively extract LCC information, and provide accurate reference for coral reef ecosystem health analysis and protection management.
- Continuous and dynamic monitoring of LCC in the Hawaii Reef (Kure Reef) and an unidentified reef in the Great Barrier Reef was conducted in 2015, 2020 and 2022 using both domestic and international satellite data. The results show that the LCC of the reefs in these two typical areas reduced.

- **Discussion and Outlook**

LCC is considered to be the most important index for measuring coral reef health, and remote sensing inversion of LCC is an important part of coral reef remote sensing monitoring. LCC can reflect the health status of the coral reef ecosystem. Global climate change, the outbreak of crown-of-thorns starfish and storm surge may cause changes in the health status of the coral reef ecosystem, and lead to changes in LCC. The inversion of LCC using multi-spectral satellite remote sensing images has good accuracy, and the model exhibits excellent scalability in terms of time, space, and sensors. However, the spectrum of coral reefs is different in different periods of time. Therefore, we need to develop advanced remote sensing monitoring methods for coral reef ecology to meet the monitoring needs of typical coral reef ecosystems.

China has successfully launched a series of marine remote sensing satellites covering ocean color, power, monitoring and so on, which provide a good opportunity for the research and application of coral reef ecological remote sensing. However, the existing remote sensing monitoring methods do not have corresponding spectral segments in coral reef research, so we are considering using the multi-source data fusion method for inversion research. With the close combination of ecology and remote sensing, the application of remote sensing monitoring of coral reefs can be systematically developed to provide information support for global change research.

7.4 Summary

This chapter focuses on two main themes: reducing marine pollution and protecting marine ecosystems. It includes case studies on the distribution and migration characteristics of microplastics in the Arctic-Pacific sector, the spatiotemporal changes of large-scale phytoplankton blooms in the North Indian Ocean, and dynamic monitoring of LCC on typical islands and reefs. The related technologies, methods, and datasets lay a foundation for better promoting the achievement of the SDGs in key global regions.

Based on the research in this chapter, we propose the following recommendations.

(1) Review and adjust the relevant indicator system using the opportunity of the 2023 midterm assessment of the SDGs. Currently, five specific targets under SDG 14 (SDG 14.1, SDG 14.2, SDG 14.3, SDG 14.a, SDG 14.c) are in a "have methods but lack data" state, which hinders the UN and national governments from effectively understanding the current status and implementing related management policies. It is necessary to adjust the existing indicator system in line with current concerns and the midterm assessment, and gather more monitoring data for SDG evaluation. For example, in SDG 14.1, additional indicators for microplastic pollution and marine radioactive contamination should be included. In SDG 14.2, evaluation indicators for the restoration of typical marine ecosystems (such as mangroves, seagrass beds, coral reefs, etc.) should be added.
(2) Encourage and guide countries to increase support for marine scientific research through the implementation of the "UN Decade of Ocean Science for Sustainable Development" program. This includes increasing investment in marine science and technology, establishing new marine science research platforms, encouraging participation from enterprises and the public, and improving the capacity for marine data collection and analysis. The program will also strengthen international cooperation in the marine field, promote the management, open access, and sharing of marine data resources, in order to better support the achievement of the marine SDGs.

In the future, we will continue to enhance the sharing and application capabilities of Big Earth Data in the field of marine sustainable development. Through building data-sharing platforms, online computing platforms, and data service platforms, we aim to facilitate the timely sharing and dissemination of data and knowledge, improve the development of the blue economy and marine technological innovation, and contribute to the implementation of the 2030 Agenda.

References

Ahn YH, Shanmugam P (2006) Detecting the red tide algal blooms from satellite ocean color observations in optically complex Northeast-Asia Coastal waters [J]. Remote Sens Environ 103(4):419–437
Amin R, Zhou J, Gilerson A et al (2009) Novel optical techniques for detecting and classifying toxic dinoflagellate *Karenia brevis* blooms using satellite imagery [J]. Opt Express 17(11):9126–9144
Borrelle SB, Ringma J, Law KL et al (2020) Predicted growth in plastic waste exceeds efforts to mitigate plastic pollution [J]. Science 369(6510):1515–1518
Cózar A, Martí E, Duarte CM et al (2017) The Arctic Ocean as a dead end for floating plastics in the North Atlantic branch of the Thermohaline Circulation [J]. Sci Adv 3(4):e1600582
Dai YH, Yang SB, Zhao D et al (2023) Coastal phytoplankton blooms expand and intensify in the 21st century [J]. Nature 615(7951):280–284
Plastics Europe (2023) Plastics-the fast facts 2023 [EB/OL]. https://plasticseurope.org/wp-content/uploads/2023/10/Plasticsthefastfacts2023-1.pdf [2024–04–07]
Fang C, Zheng RH, Zhang YS et al (2018) Microplastic contamination in benthic organisms from the Arctic and sub-Arctic regions [J]. Chemosphere 209:298–306

Fang C, Zheng RH, Hong FK et al (2021) Microplastics in three typical benthic species from the Arctic: Occurrence, characteristics, sources, and environmental implications [J]. Environ Res 192:110326

Hu CM, Feng L (2017) Modified MODIS fluorescence line height data product to improve image interpretation for red tide monitoring in the eastern Gulf of *Mexico* [J]. J Appl Remote Sens 11(1):012003

Hu CM, Muller-Karger FE, Taylor C et al (2005) Red tide detection and tracing using MODIS fluorescence data: a regional example in SW *Florida* coastal waters [J]. Remote Sens Environ 97(3):311–321

Kanhai LDK, Johansson C, Frias JPGL et al (2019) Deep sea sediments of the Arctic central basin: a potential sink for microplastics [J]. Deep Sea Res Part I 145:137–142

Liao ZH, Yu KF, Chen B et al (2021) Spatial distribution of benthic algae in the South China Sea: responses to gradually changing environmental factors and ecological impacts on coral communities [J]. Divers Distrib 27(5):929–943

Lima ARA, Ferreira GVB, Barrows APW et al (2021) Global patterns for the spatial distribution of floating microfibers: Arctic Ocean as a potential accumulation zone [J]. J Hazard Mater 403:123796

Liu RJ, Xiao YF, Ma Y et al (2022) Red tide detection based on high spatial resolution broad band optical satellite data [J]. ISPRS J Photogramm Remote Sens 184:131–147

Lu WF, Gao XY, Wu ZL et al (2022) Framework to extract extreme phytoplankton bloom events with remote sensing datasets: a case study [J]. Remote Sensing 14(15):3557

Mu JL, Qu L, Jin F et al (2019) Abundance and distribution of microplastics in the surface sediments from the northern Bering and Chukchi Seas [J]. Environ Pollut 245:122–130

Obbard RW, Sadri S, Wong YQ et al (2014) Global warming releases microplastic legacy frozen in Arctic Sea ice [J]. Earth's Futur 2(6):315–320

Peeken I, Primpke S, Beyer B et al (2018) Arctic sea ice is an important temporal sink and means of transport for microplastic [J]. Nat Commun 9(1):1505

Polyakov IV, Pnyushkov AV, Timokhov LA (2012) Warming of the intermediate Atlantic water of the arctic ocean in the 2000s [J]. J Clim 25(23):8362–8370

Ross PS, Chastain S, Vassilenko E et al (2021) Pervasive distribution of polyester fibres in the arctic ocean is driven by Atlantic inputs [J]. Nat Commun 12(1):106

Rudels B (2001) Arctic basin circulation [M]. In: Steele JH, Thorpe SA, Turekian KK (eds) Encyclopedia of ocean sciences. San Diego: Academic Press: 177–187

Souter D, Planes S, Wicquart J, et al (2021) Status of coral reefs of the world: 2020 [R]. https://gcrmn.net/wp-content/uploads/2023/04/GCRMN_Souter_et_al_2021_Status_of_Coral_Reefs_of_the_World_2020_V1.pdf [2024–10–10]

United Nations (2023b) Progress towards the sustainable development goals: towards a rescue plan for people and planet, report of the secretary-general (special edition) [R]. https://digitallibrary.un.org/nanna/record/4014344/files/A_78_80--E_2023_64-EN.pdf?withWatermark=0&withMetadata=0&version=1&registerDownload=1 [2023–04–27]

Wang MH, Jiang LD, Mikelsons K et al (2021) Satellite-derived global chlorophyll-a anomaly products [J]. Int J Appl Earth Obs Geoinf 97:102288

Zhao MX, Yu KF, Shi Q et al (2013) Coral communities of the remote atoll reefs in the Nansha Islands, southern South China Sea [J]. Environ Monit Assess 185(9):7381–7392

8 SDG 15 Life on Land

8.1 Background

SDG 15 focuses on sustainably managing forests, combating desertification, halting and reversing land degradation, and halting biodiversity loss. Eight years after the adoption of the 2030 Agenda, we are still facing a critical situation, as the global forest area (SDG 15.1.1) continues to steadily decline (FAO and UNEP 2020), and approximately 75% of the world's land is still degrading (SDG 15.3.1) (IPBES 2019). The proportion of globally important protected biodiversity sites has increased (SDG 15.1.2 and SDG 15.4.1), but the Red List Index (SDG 15.5.1) continues to decline (UNEP 2021a), and at the current rate SDG 15 will be difficult to achieve by 2030 (United Nations 2019).

Assessing progress toward SDG 15 is key to identifying gaps and implementing effective interventions. With improved data availability and technical methodologies, eight of the fourteen indicators covered by SDG 15 are in Tier I (with methods and data). However, the methods for obtaining these indicators are mainly based on statistical instruments, which lack scalability (global-regional-national-local), and many countries with limited data availability are unable to provide data on a regular basis. Therefore, it is necessary to further develop multiscale and spatialized research on key technologies for monitoring the status and progress of SDG 15 indicators using cutting-edge technologies, such as Earth observation, artificial intelligence, and public science, to contribute from the perspectives of data, methods, tools, and decision-making recommendations, and thus provide scientific and technological support for the realization of SDG 15.

This chapter will focus on three directions: land desertification, grassland degradation, and migratory bird risk assessment. It focuses on specific objectives such as land degradation neutrality (SDG 15.3), and habitat protection of endangered species (SDG 15.5). The goal is to dynamically monitor and evaluate the sustainable development process of terrestrial organisms through Big Earth Data technology in the entire area or typical areas of the region involved in the BRI, and provide scientific and technological support for the monitoring and evaluation of SDG 15 indicators.

8.2 Main Contributions

To address the data gaps in SDG 15 progress assessment, we produced high-precision mapping of desertification on the Mongolian Plateau in 2015 and 2020, the 2000–2020 global grassland degradation and improvement distribution dataset, the global grassland degradation and improvement degree dataset, the global grassland degradation and improvement driving force

H. Guo, *Big Earth Data in Support of the Sustainable Development Goals (2023)—The Belt and Road*, Sustainable Development Goals Series, https://doi.org/10.1007/978-981-95-3178-3_8

Table 8.1 Cases and their main contributions

Targets	Cases	Contributions
SDG 15.3 By 2030, combat desertification, restore degraded land and soil, including land affected by desertification, drought and floods, and strive to achieve a land degradation-neutral world	Evolution of desertification and control strategies in the Mongolian plateau (2015–2020)	Data product: High-precision mapping of desertification on the Mongolian Plateau in 2015 and 2020
	Global grassland degradation trends from 2000 to 2020	Data product: 2000–2020 global grassland degradation and improvement distribution dataset, the global grassland degradation and improvement degree dataset, and the global grassland degradation and improvement driving force dataset Method and model: Using remote sensing evaluation technology that couples vegetation productivity with climate factors, precise monitoring of global grassland degradation trends from 2000 to 2020 was achieved Decision support: This contributes to curbing grassland degradation and promoting ecological restoration, providing a scientific basis for formulating sustainable development strategies
SDG 15.5 take urgent and significant action to reduce the degradation of natural habitats, halt the loss of biodiversity and, by 2020, protect and prevent the extinction of threatened species	Risk assessment of migratory birds from nighttime light expansion in major cities around the world	Data product: Spatial distribution map of migratory bird richness in major cities around the world, spatiotemporal variation map of NTL expansion, and the distribution map of the exposure level of migratory birds in major cities around the world Decision support: Provide support for migratory bird conservation

dataset, spatial distribution map of migratory bird richness in major cities around the world, spatiotemporal variation map of NTL expansion, and the distribution map of the exposure level of migratory birds in major cities around the world, which can provide important support for the assessment of terrestrial biodiversity SDGs at regional and global scales (Table 8.1).

8.3 Case Study

8.3.1 Evolution of Desertification and Control Strategies in the Mongolian Plateau (2015–2020)

Target: SDG 15.3 by 2030, combat desertification, restore degraded land and soil, including land affected by desertification, drought and floods, and strive to achieve a land degradation-neutral world.

- **Background**

The Mongolian Plateau is the core area of the China-Mongolia-Russia Economic Corridor along the BRI. At the 22nd meeting of the Council of Heads of Member States of the Shanghai Cooperation Organization in 2022, the heads of State of China, Mongolia and Russia stated that the development plan of the China-Mongolia-Russia Economic Corridor Would be extended for five years. As one of the most significant arid regions in the Northern Hemisphere, the Mongolian Plateau is a sensitive and vulnerable area in terms of global change response, also playing an important role in the construction of an ecological security barrier in the north of the country and in Northern Asia as a whole. Desertification is one of the most serious ecological and environmental problems in the region, and also a challenge for the realization of SDG 15.3. Under the combined influence of climate change and irrational human activities such as overgrazing and uncontrolled mining, the problems of grassland degradation and soil desertification in the region have become increasingly serious. Mongolia is a hotspot for global desertification. According to the Ministry of Environment and Tourism of Mongolia, in 2015 and 2017, the total area of desertified and degraded land in Mongolia accounted for 76.9% and 76.8% of the country's total land area, respectively. Nei Mongol Autonomous Region is one of the provinces (autonomous regions, municipalities) in China with the largest desertified land area. According to the Fifth National Monitoring of Desertification and Sandification, the desertified land area in Nei Mongol reached 609,200 km^2 in 2014, accounting for 50.12% of its total land area (Yang 2020). To accurately assess the realization of the land degradation neutrality goal, high-resolution data products on land degradation and desertification are essential. However, due to the limitations of traditional desertification monitoring methods and technical approaches, there is a lack of long time-series and highly detailed desertification data products covering the whole area of the Mongolian Plateau. Therefore, the distribution of desertification degree and the fine evolution of desertification in the whole Mongolian Plateau have not been fully grasped, as well as the fine differences in desertification evolution under the influence of different grazing systems and ecological protection policies in China and Mongolia.

- **Data**

  - Landsat8 remote sensing image data. The data comes from the USGS website (http://earthexplorer.usgs.gov/), with a spatial resolution of 30 m from July to September in 2015 and 2020.
  - Meteorological data: A monthly-scale, high spatial resolution TerraClimate climate dataset is used with a spatial resolution of about 4 km, including meteorological factors such as precipitation, maximum temperature, minimum temperature, saturated vapor pressure, wind speed, potential evapotranspiration and soil moisture.
  - Socio-economic data: including population and livestock carrying capacity data from 2015 to 2020, sourced from the National Statistics Office of Mongolia and Bureau of Staistics of Nei Mong Autonomous Region.
  - Auxiliary data: DEM data, land cover data, etc. of the Mongolian Plateau.

- **Method**

In order to adapt to the characteristics of the Mongolian Plateau, this case study identified the potential extent of desertification on the Mongolian Plateau. An optimal indicator system and method suitable for monitoring desertification on the Mongolian Plateau were constructed. The long-term lack of fine desertification data in the region was solved. The contribution of meteorological and anthropogenic factors to the spatial distribution and evolution of desertification in the Mongolian Plateau was clarified. This provides scientific decision support for desertification prevention and control in this region and other large arid and semi-arid areas in the world.

Firstly, based on historical literature research and field surveys, a desertification grading indicator system was constructed for the Mongolian Plateau, which was divided into five grades, namely, severe, heavy, moderate, light and potential desertification. According to the desertification definition of the United Nations Convention to Combat Desertification, the potential range of desertification occurring on the Mongolian Plateau was delineated based on precipitation and potential evapotranspiration data. Subsequently, all Landsat8 images of the growing season (June–September) were filtered on GEE with cloud masking and quality screening, and normalized difference vegetation index (NDVI) maximum synthesis was used to obtain the images with the best annual vegetation growth conditions; the synthesized images were used to calculate vegetation and soil indicators, and combined with Landsat original spectral information and topographic data to map the desertification degree distribution using machine learning methods in 2015 and 2020. Finally, the spatial and temporal evolution of desertification on the Mongolian Plateau was analyzed, and the driving mechanisms of desertification at different regional scales on the Mongolian Plateau were further explored using Geodetector models.

- **Results and Analysis**

1. **Grading System for Desertification on the Mongolian Plateau**

In order to meet the needs of desertification assessment and mapping, in 1984, the FAO and the United Nations Environment Programme (UNEP) divided desertification into five grades: potential, light, moderate, heavy, and severe. Based on the historical documents of the Mongolian Plateau and combined with field data, this case study constructed a grading system for desertification degree suitable for the Mongolian Plateau (Table 8.2). Landscape photos from the Selenge River Basin field trip were taken in 2023.

2. **Construction of a Desertification Sample Database for the Mongolian Plateau**

Following the established desertification classification standards, 1,000 desertification sample

Table 8.2 Grading criteria for desertification degree on the Mongolian Plateau

Desertification grades	Landscape photos	Surface features	Vegetation cover/%
Potential		Dominated by forests, cropland and high-cover grasslands	>65
Light		Vegetation begins to decline, patches of sand and poisonous weeds appear, and the area of degraded land is 5–25%	(50, 65]
Moderate		Degraded vegetation occurs and is distributed with low scrub sand piles, and degraded land area accounts for 26–50% of the land area	(10, 50]
Heavy		Localized loss of vegetation cover, mixing of sandy pioneer plants with the original plant populations of the pasture, and more than 50% of the degraded land area	(1, 10]
Severe		The land is completely unproductive and there is a regional loss of vegetation, mainly bare land, sand, Gobi, etc	<1

Note The images in Table 8.2 were photographed by the author(s)

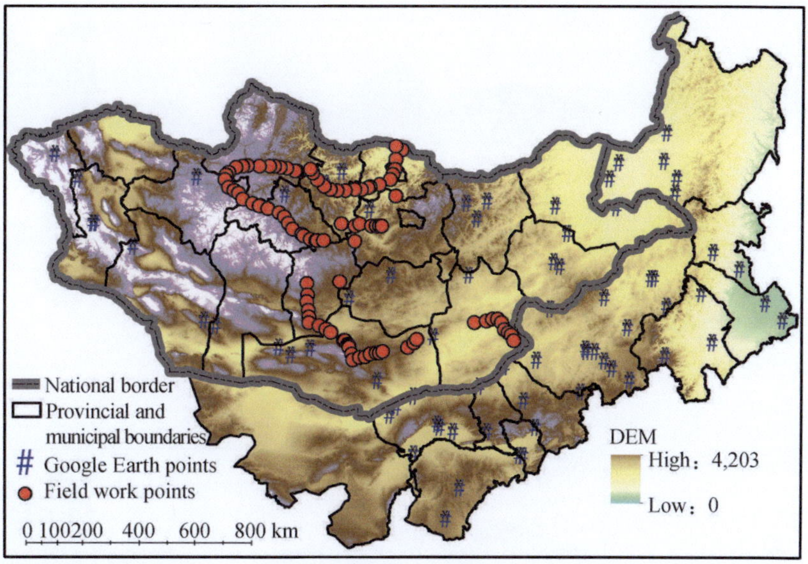

Fig. 8.1 The distribution of desertification sample points on the Mongolian Plateau

points were selected using high-resolution Google Earth imagery and Sentinel satellite imagery with the assistance of land cover data, vegetation coverage data, and other materials on the Mongolian Plateau (Fig. 8.1).

3. **Spatial Distribution and Evolution of Desertification Degree on the Mongolian Plateau**

The potential and light desertification were mainly concentrated in the northern and eastern regions of the Mongolian Plateau, including the provinces of Khentii, Dornod, Sükhbaatar, Töv, Bulgan, and Uvurkhangai in northern Mongolia, and Hulunbuir City, Xing'anmeng, and Tongliao City in Nei Mongol, China. The heavy and severe desertification were mainly found in the southwestern part of the Mongolian Plateau including the Ömnögovi Province, Dornogovi Province, and Dundgovi Province, and Arkhangai Province in central and southern Mongolia, and the Bayannur City, Alshaa League, Baotou City, and Ulanqab citiy in northern Nei Mongol, China. Moderate desertification was mainly located within the transition zone between light and heavy desertification. As the area of potentially and lightly desertified land in Northern Mongolia and Southeastern Nei Mongol further expanded from 2015 to 2020, the area of severely desertified land in Southeastern Mongolia also expanded. In 2015, the desertified land area of the Mongolian Plateau reached 1.698×10^6 km^2, and in 2020, the desertified land area decreased to 1.656×10^6 km^2. Among them, the area of severely and lightly desertified land increased, and the area of other desertified land types decreased. At the same time, the area of desertified land in China's Nei Mongol and Mongolia decreased. However, the area of severely desertified land in Mongolia increased significantly, from 1.964×10^5 km^2 in 2015 to 2.445×10^5 km^2, and the area of severely desertified land in China's Nei Mongol decreased by about 10^5 km^2 from 2015 to 2020 (Fig. 8.2).

From 2015 to 2020, the overall desertified land in the Mongolian Plateau was restored. The central and western parts of the Mongolian Plateau were the main areas of land deterioration, mainly in the southern and western parts of Mongolia, such as Bayanhongor, Ömnögovi Province, Govi-Altai, Dundgovi Province and Arkhangai Province. By contrast, Nei Mongol has basically entered a state of full recovery. For the Mongolian Plateau, Nei Mongol and Mongolia, the evolution of moderate desertification to heavy desertification and heavy desertification to severe desertification is the main mode of degradation, and the transition from

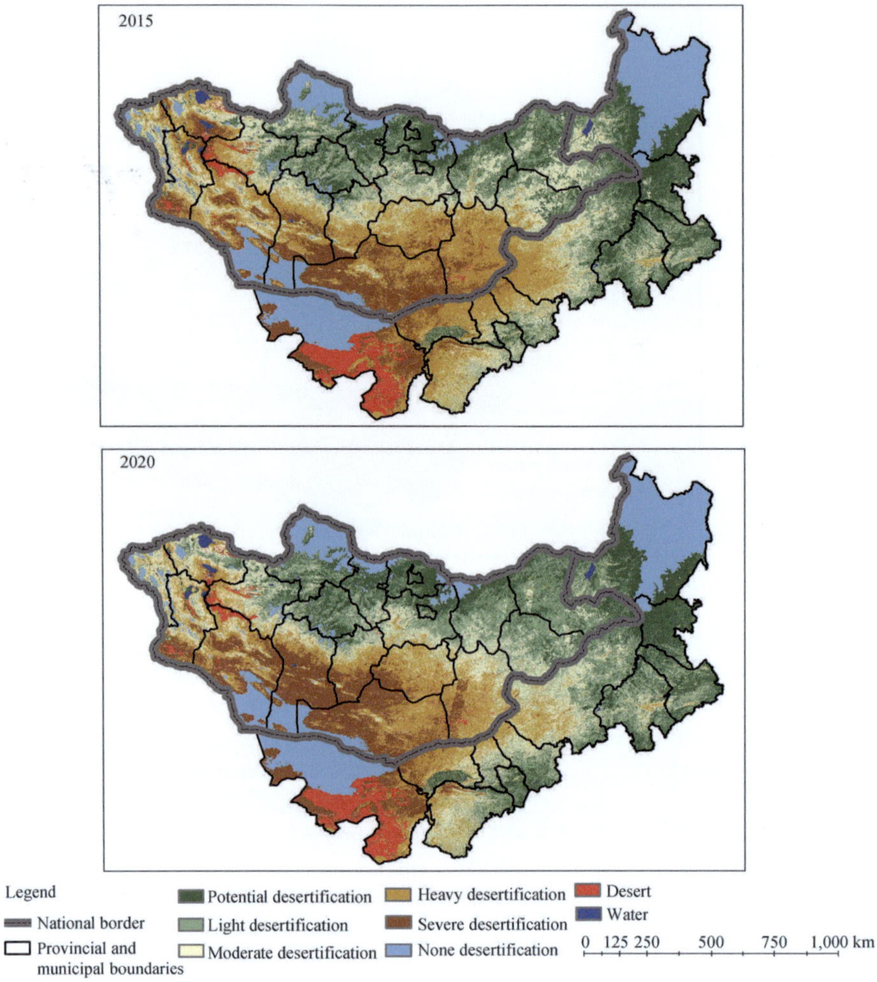

Fig. 8.2 The distribution of desertification on the Mongolian Plateau in 2015 and 2020

potential desertification to moderate desertification is the main mode of significant degradation. The transition from severe desertification to heavy desertification, from heavy desertification to moderate desertification, and from moderate desertification to light desertification are the main modes of restoration (Fig. 8.3).

Recovery and deterioration of desertified land depend on climate change and the intensity of human activities. Precipitation and land use are the main drivers of the spatial distribution pattern of desertification in the Mongolian Plateau and Mongolia. Potential evapotranspiration and precipitation are the main influencing factors in Nei Mongol. Land use change is the main driver of desertification evolution in the Mongolian Plateau and Mongolia. Since the 1980s, the average annual temperature in Mongolia has risen by about 1.5–2.5 °C, which is 2–4 times higher than the global average temperature. A significant increase in temperature is likely to exacerbate the frequency of droughts and heat waves, leading to further deterioration of desertified land. From 2015 to 2020, the increasing trend of precipitation promoted vegetation growth. Overgrazing has further exacerbated the deterioration of desertified land in Mongolia. Since the early 1990s, Mongolia has been practicing a planned economy and animal husbandry policy, that is, "pastures are owned by pastoralists" and "livestock is privately

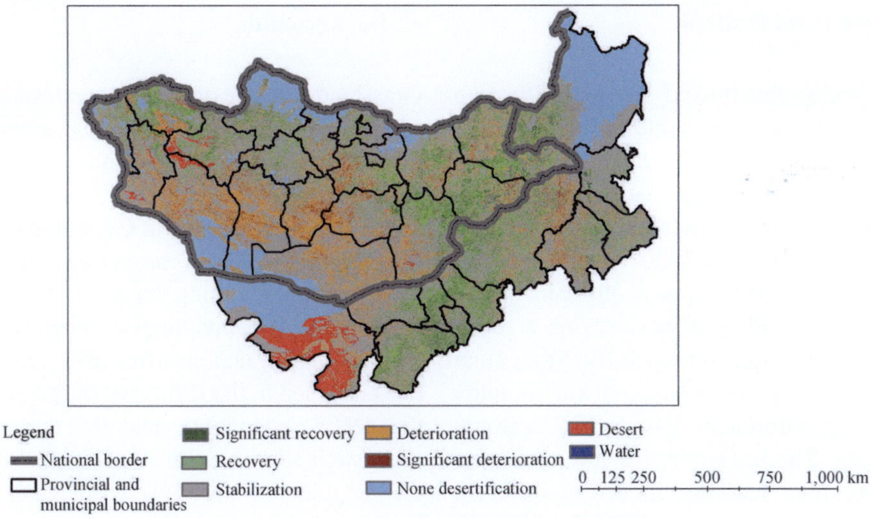

Fig. 8.3 Evolution of desertification in the Mongolian Plateau from 2015 to 2020

owned". In order to obtain the right to use pastures and pursue economic benefits, pastoralists compete to increase the number of livestock and raise goats in large numbers. From 1990 to 2020, the total number of livestock increased by about 41.75 million heads, or 1.6 times; the total number of goats increased by about 22.6 million, or 4.41 times. The rapid increase in the total number of livestock and the imbalance in the structure of livestock have exacerbated grassland degradation and land desertification. To combat desertification, Mongolia has enacted a series of laws and regulations and ecological restoration projects, including the National Program to Combat Desertification and the National "Great Green Wall" Program. While these measures are expected to play a key role in combating desertification, they have had little effect due to limited funding and lack of continuity. At the same time, grazing intensity in China's Nei Mongol is also increasing, with livestock numbers increasing by 33.48 million heads between 2000 and 2010, but have remained stable since then. However, unlike the nomadic way of Mongolian herding, Nei Mongol has implemented a strict strategy of enclosure and rotational grazing since 2000, namely grazing prohibition, fallow grazing, rotational grazing and house feeding. This strategy has promoted the fundamental transformation of the mode of animal husbandry production and management and the strategic adjustment of the economic structure, and has reduced the pressure on the utilization of grassland resources.

Highlights

- The potential scope of desertification on the Mongolian Plateau was determined, the optimal indicator system and method suitable for monitoring desertification on the Mongolian Plateau was constructed, and high-precision mapping of desertification on the Mongolian Plateau in 2015 and 2020 was completed.
- The spatiotemporal distribution pattern and evolution process of desertification on the Mongolian Plateau were revealed, and the differences in desertification evolution between China and Mongolia were further compared and analyzed.
- The contribution of meteorological factors and anthropogenic factors to the spatial distribution and evolution of desertification in different regions of the Mongolian Plateau was clarified, providing scientific decision support for desertification prevention and control in this region and other large arid and semi-arid regions in the world.

- **Discussion and Outlook**

This case study constructed the desertification degree grading system suitable for the Mongolian Plateau and the sample database of desertification on the Mongolian Plateau. The spatial distribution map of desertification on the Mongolian Plateau in 2015 and 2020 was drawn. The distribution of desertification at different levels on the Mongolian Plateau can provide a reference for desertification control on the Mongolian Plateau. The Mongolian Plateau has a complex geographical environment and obvious regional heterogeneity. The uncertainty of the results of a single machine learning model in different internal geographical units is high. In addition, due to the lack of soil attribute data such as soil organic carbon and climate data with high temporal and spatial resolution on the Mongolian Plateau, a small amount of vegetation and soil indicators obtained by remote sensing image inversion can reduce but cannot completely avoid the impact of long-term climate fluctuations on the extraction of refined desertification information and the succession process in the region, especially in the transition zone between arid and semiarid areas. Therefore, future research should focus on the combination of different geographical units and machine learning models, and further explore remote sensing inversion indicators that can reflect soil properties and climate characteristics, as well as soil properties and climate products with high temporal and spatial resolution.

8.3.2 Global Grassland Degradation Trends from 2000 to 2020

Target: SDG 15.3 By 2030, combat desertification, restore degraded land and soil, including land affected by desertification, drought and floods, and strive to achieve a land degradation-neutral world.

- **Background**

Grasslands are one of the most widespread terrestrial vegetation types, covering approximately one-fifth of the global land area. They serve as fundamental land resources essential for human survival and development. Grassland ecosystems play critical roles in providing food, sequestering carbon, releasing oxygen, conserving water resources, and maintaining biodiversity. However, climate change and unsustainable land use practices have led to the shrinkage of grassland areas, the decline in quality, and the degradation of ecosystem services. These phenomena of grassland degradation pose severe threats to grassland health and human well-being. In recent years, the rapid advancement of remote sensing technology has provided powerful support for monitoring grassland degradation on a global scale. Therefore, conducting a comprehensive assessment of global grassland degradation, analyzing its spatiotemporal dynamics, and understanding its responses to climate change and human activities are crucial for understanding the current status of grassland degradation and improvement worldwide. Such efforts provide fundamental data and practical recommendations for restoring degraded grasslands. Furthermore, scientifically evaluating global grassland degradation trends is significant for achieving SDG 15.3, which aims to create "a land degradation-neutral world".

- **Data**

  - Vegetation productivity data: The global vegetation NPP dataset for 2000–2020, with a spatial resolution of 500 m (Yuan et al. 2010).
  - Meteorological data: The global atmospheric reanalysis dataset for 2000–2020, with a spatial resolution of 0.125°; The Global Precipitation Climatology Centre (GPCC) global precipitation reanalysis

dataset for 2000–2013, with a spatial resolution of 0.5°; The MERRA-2 global meteorological reanalysis dataset for 2014–2020, with a spatial resolution of 0.50° × 0.625°.

- Global land cover data: The ESA 2015 GlobCover dataset, with a spatial resolution of 300 m.

- **Method**

This case study presents a remote sensing evaluation technique for grassland degradation based on changes in vegetation productivity and climate coupling. This technique was applied to assess global grassland degradation trends from 2000 to 2020, producing global datasets for grassland degradation and improvement distribution, degree of degradation and improvement, and the driving force of degradation and improvement.

Firstly, based on the global GLASS-NPP long-term time series data product and concurrent meteorological data, this case study developed an indicator system for assessing grassland degradation. This system incorporates NPP (representing vegetation productivity), the moisture index (MI) (representing interannual climate wetness and dryness), and a newly defined NPP climate response rate—moisture-responded net primary productivity (MNPP) (representing whether vegetation growth changes reflect responses to interannual climate variability).

Secondly, the Sen + Mann–Kendall trend test method was employed, combined with the response relationship between MNPP and MI, to determine and assess the spatial extent of global grassland degradation and improvement.

Thirdly, considering the characteristics of grassland degradation and improvement, and based on expert knowledge, the NPP rate of change was used to assess the degree of global grassland degradation and improvement.

Lastly, a driving force model for grassland degradation was developed using multiple stepwise regression and partial regression methods. This model quantifies the impact of climate and human factors, distinguishing between the combined effects of climate and human factors, climate-only factors, and human-only factors on grassland degradation trends, thereby enabling the quantitative identification of the driving forces behind grassland degradation.

- **Results and Analysis**

The global grassland degradation trends from 2000 to 2020 are shown in Fig. 8.4. Statistical analysis indicates that the global grassland degradation trends recovered overall during this period. The area of improved grassland is approximately 1.9 times that of degraded grassland, and there are specific spatial differences in the distribution of these two trends. From 2000 to 2020, about 1.938 million km^2 of global grassland experienced intensified degradation, primarily concentrated in Ethiopia and southern Sudan, Kenya, central and southern Republic of the Congo and southern Democratic Republic of the Congo, western and southern Angola and Namibia, the Drakensberg Mountains, and southern Madagascar in Africa; the Brazilian Plateau, the Orinoco Plain, and northern Bolivia in South America; central Australia; and the northern part of the Caspian Sea in Asia. Areas of global grassland improvement and restoration totaled approximately 3.732 million km^2, with spatial concentrations in northern Mongolia, eastern Russia, the Nei Mongol Plateau, the Loess Plateau and surrounding areas, Xizang, the Deccan Plateau in India, as well as the border regions between Pakistan and Afghanistan in Asia; the Missouri River Basin in the central Great Plains of North America, northern Canada, and parts of Alaska; the central-western Sahel and northern Namibia in Africa; and the Paraguay River Basin and southern Peru in South America. Apart from these areas, there are still scattered regions across other continents where grassland degradation and improvement are observed, though the affected areas are smaller and more dispersed.

The trends of global grassland degradation and improvement at varying degrees from 2000 to 2020 are shown in Fig. 8.5. Among the degraded grasslands, more than 75% experienced moderate

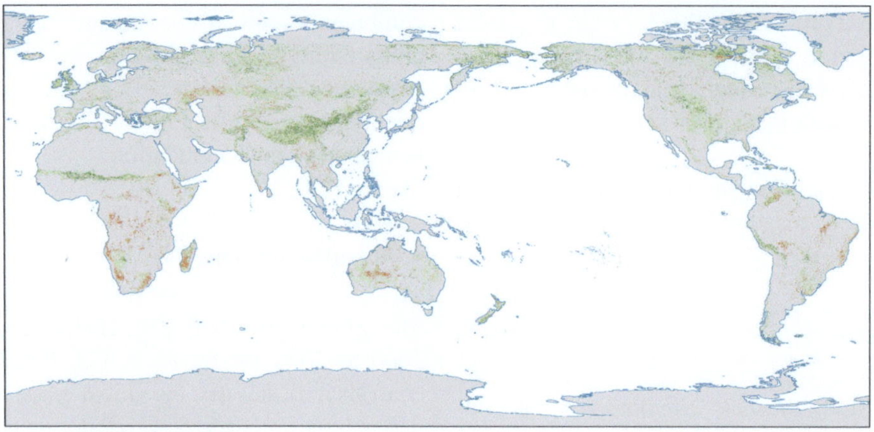

Fig. 8.4 Global grassland degradation trends from 2000 to 2020

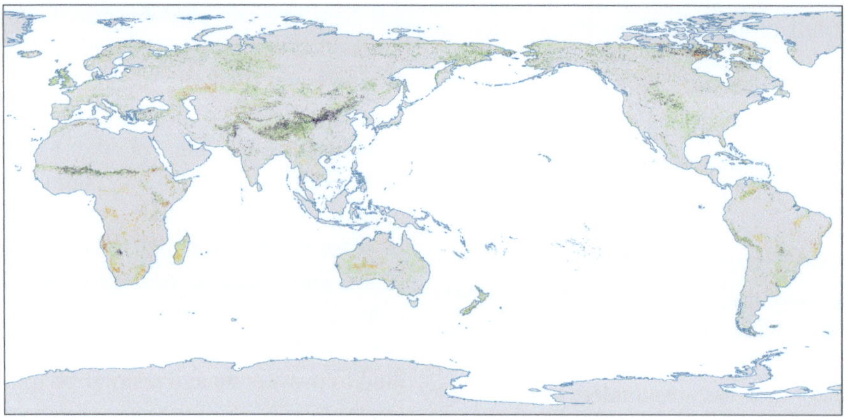

Fig. 8.5 Global grassland degradation and improvement trends in varying degrees from 2000 to 2020

or slight degradation, while the areas of significant degradation and very significant degradation were similar, measuring 235,300 km^2 and 229,600 km^2, respectively. The very significant degradation and significant degradation were primarily concentrated in northern Canada, the northern part of the Caspian Sea in Asia, and parts of Africa, including Kenya, southern Sudan, Angola, and Namibia. In other regions with concentrated grassland degradation, most areas experienced moderate or slight intensified degradation. Among the improved grasslands, the area that underwent moderate significant improvement was the largest, approximately 1.3238 million km^2, accounting for about 35% of the total improved grassland area. The area of very significantly improved grasslands ranked second, covering 910,500 km^2, followed by slightly and significantly improved grasslands. The distribution of very significantly and significantly improved grasslands was most notable in China's Nei Mongol Plateau, Loess Plateau, and their surrounding areas. Other concentrated areas included northern and central North America, the central-western Sahel region of Africa, northern Namibia, and the border regions between Afghanistan and Pakistan. Moderately

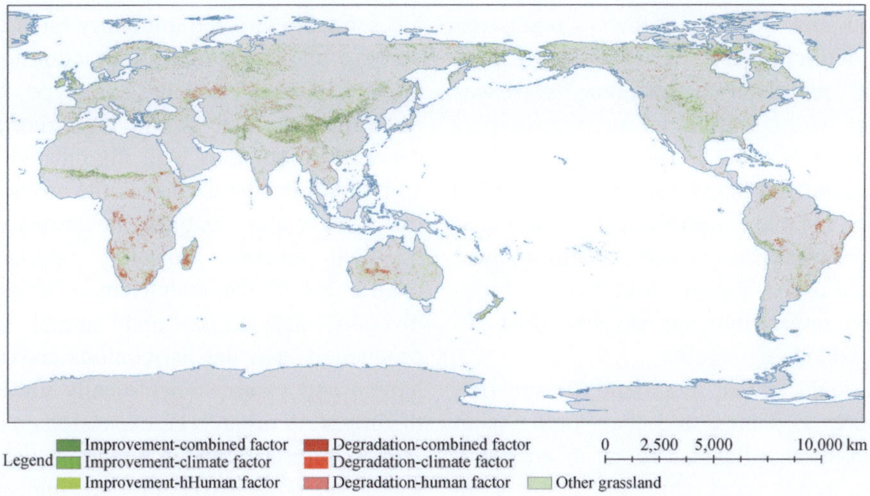

Fig. 8.6 Driving forces of global grassland degradation and improvement trends from 2000 to 2020

and slightly improved grasslands were mainly distributed in Xizang, southern Peru, northern Alaska in North America, and eastern Russia, while these improvement trends also dominated other scattered areas of improved grasslands.

The driving forces of global grassland degradation and improvement trends from 2000 to 2020 are shown in Fig. 8.6. Statistical analysis reveals that global grassland degradation was primarily driven by a combination of climate and human factors, affecting 84.7% of the total degraded grassland area, with distribution across the most degraded regions. Grassland degradation driven solely by human factors covered 284,200 km^2, with notable concentrations in Namibia in Africa and the northern part of the Caspian Sea in Asia. Grassland degradation driven solely by climate factors accounted for the smallest area, only 0.63% of the total degraded area. Similarly, global grassland improvement was predominantly influenced by the combined effects of climate and human factors, affecting over 3.25 million km^2, or 87.76% of the total improved grassland area, and was widely distributed across improvement regions. Grassland improvement driven solely by human factors covered 441,900 km^2 and was scattered throughout the improvement regions. Grassland improvement driven solely by climate factors accounted for the smallest area, representing only 0.4% of the improved area.

Highlights

- This case study produced the 2000–2020 global grassland degradation and improvement distribution dataset, the degree of global grassland degradation and improvement dataset, and the driving force of global grassland degradation and improvement dataset. These datasets provide essential data support for accurately understanding the area, degree, spatial distribution, and driving forces of global grassland degradation.
- From 2000 to 2020, global grassland degradation and improvement processes proceeded in parallel across different regions. However, the area of improved grasslands was approximately 1.9 times larger than the area of degraded grasslands, indicating a clear trend of global grassland recovery. Notably, over 40% of grasslands showed significant improvement. Climate and human factors have also played key roles in driving the degradation and improvement processes of grasslands.

- **Discussion and Outlook**

A comprehensive evaluation of global grassland degradation trends and driving forces was conducted using the Sen + Mann–Kendall trend test method and multiple stepwise regression based on long-term NPP, MNPP, and MI data. The global grassland degradation map was produced, providing critical information support for achieving SDG 15.3.1. From 2000 to 2020, global grasslands exhibited a significant recovery trend, with notable improvements. The dual driving forces of climate and human factors played a crucial role in the improvement process, highlighting the remarkable success of efforts to combat grassland degradation.

In traditional global grassland degradation regions, such as China and the South Asian subcontinent, a significant reversal in grassland degradation trends were observed. Notably, in China's Nei Mongol Plateau, and the Loess Plateau and its surrounding areas, a series of ecological projects, including programs like converting farmland to grassland and returning grazing land to grassland, have significantly improved vegetation cover in degraded grasslands, leading to their restoration and recovery. This demonstrates that practical policy guidance and proactive, scientifically informed management measures can reverse grassland degradation trends successfully. Asia serves as a valuable example, providing actionable insights for global grassland degradation prevention and the achievement of SDG 15.3.1.

8.3.3 Risk Assessment of Migratory Birds from Nighttime Light Expansion in Major Cities Around the World

Target: SDG 15.5 Take urgent and significant action to reduce the degradation of natural habitats, halt the loss of biodiversity and, by 2020, protect and prevent the extinction of threatened species.

- **Background**

The rapid increase in artificial lighting around the world has fundamentally changed the NTL environment in cities, and this light pollution has seriously affected migratory birds due to the positive phototaxis effect (attraction to light) of birds. Light pollution has two types of effects on birds: the first is the high mortality rate caused by collisions with illuminated buildings and windows; the second is the stranding event that occurs after seabirds are attracted to land by light sources. Protecting migratory birds contributes to the maintenance of global bird diversity, agricultural and animal husbandry production, and the harmonious coexistence of human and nature. Scientifically understanding the impact of urban NTL expansion on migratory bird migration provides important data support and a scientific basis for realizing sustainable ecosystem management and the SDGs. Existing studies have mainly focused on national and region-specific scales, but research on global scales is still insufficient. The research of this case study on a global scale includes the spatial distribution map of migratory bird richness in major cities around the world, the Spatiotemporal variation map of NTL expansion, and the distribution map of the exposure level of migratory birds.

- **Data**

  – SDGSAT-1 low-light data.
  – Cornell University's multi-source geographic data for bird observations.
  – DMSP-OLS and VIIRS-DNB (visible infrared imaging radiometer suite-day/night band) NTL satellite data.

- **Method**

1. **Identification of Spatiotemporal Patterns of Migratory Birds in Major Cities Around the World**

The richness of migratory birds and the spatiotemporal distribution of species in major cities around the world were mapped by remote sensing using multi-source geographic big data for bird observation. In order to identify the spatial units of good investigation from multi-source geographic big data, it is necessary to evaluate the survey integrity of species distribution data at different spatial resolutions based on the global species diversity database, so as to obtain the optimal research scale of spatial units. Finally, the spatiotemporal distribution map of migratory birds in major cities around the world was generated based on the optimal scale.

2. **Evaluation of NTL Expansion in Major Cities Around the World**

Considering the availability, consistency and completeness of the data, the visible-near infrared satellite remote sensing data from multiple data sources was selected as the basic database for the evaluation of NTL expansion in major cities around the world, and the data were calibrated year by year and simulated in space downscaling through machine learning methods to obtain long-time, complete and consistent NTL images of major cities around the world, and generate spatiotemporal variation maps of NTL expansion in major cities around the world.

3. **Assessment of the Exposure of Migratory Birds in Major Cities Around the World**

The major cities in the world are important areas for migratory birds, and the intensity and expansion rate of urban NTL are important factors that interfere with migration, there are temporal and spatial differences in the distribution of migratory birds in each city, so it is necessary to focus on the rapid expansion cities and medium and large cities where the distribution of migratory birds is concentrated.

Based on the spatial distribution data of species richness of migratory birds in major cities around the world and the NTL data of urban expansion, the exposure levels of cities along the route were calculated using the light radiation attenuation model. This case study seeks the optimal dynamic balance between urban development and migratory bird conservation, and the evaluation results can provide important decision-making reference for relevant government departments in major cities.

- **Results and Analysis**

1. **Spatial Distribution Pattern of Migratory Bird Richness in Major Cities Around the World**

The 20 major cities in the world are located around nine migratory bird routes. The spatial distribution of migratory bird richness in major cities around the world is shown in Fig. 8.7. Among them, the richness of Los Angeles, Mumbai and Kuala Lumpur is 66, 61 and 58, respectively, which is relatively high. Cairo and Bandung have relatively low richness of 8 and 18, respectively.

2. **Spatiotemporal Variation of NTL Expansion in Major Cities Around the World**

The spatiotemporal variation of NTL expansion in major cities around the world during 1992–2022 is shown in Fig. 8.8. Our findings show that NTL in major cities around the world mostly showed an expansion trend from 1992 to 2022. Cairo, as the city with the most obvious changes, saw a 66.7% increase in NTL intensity compared with 1992, and its high-value light area expanded from southwest to northeast. However, Kabul showed a trend of gradual aggregation of large areas with high values, and the intensity of light at night increased significantly, reaching 60.5%. In addition to these two cities, some other cities, such as Hanoi, Beijing, Bandung, Caracas and other cities, also expanded significantly, with Hanoi accounting for more than half of the total area in

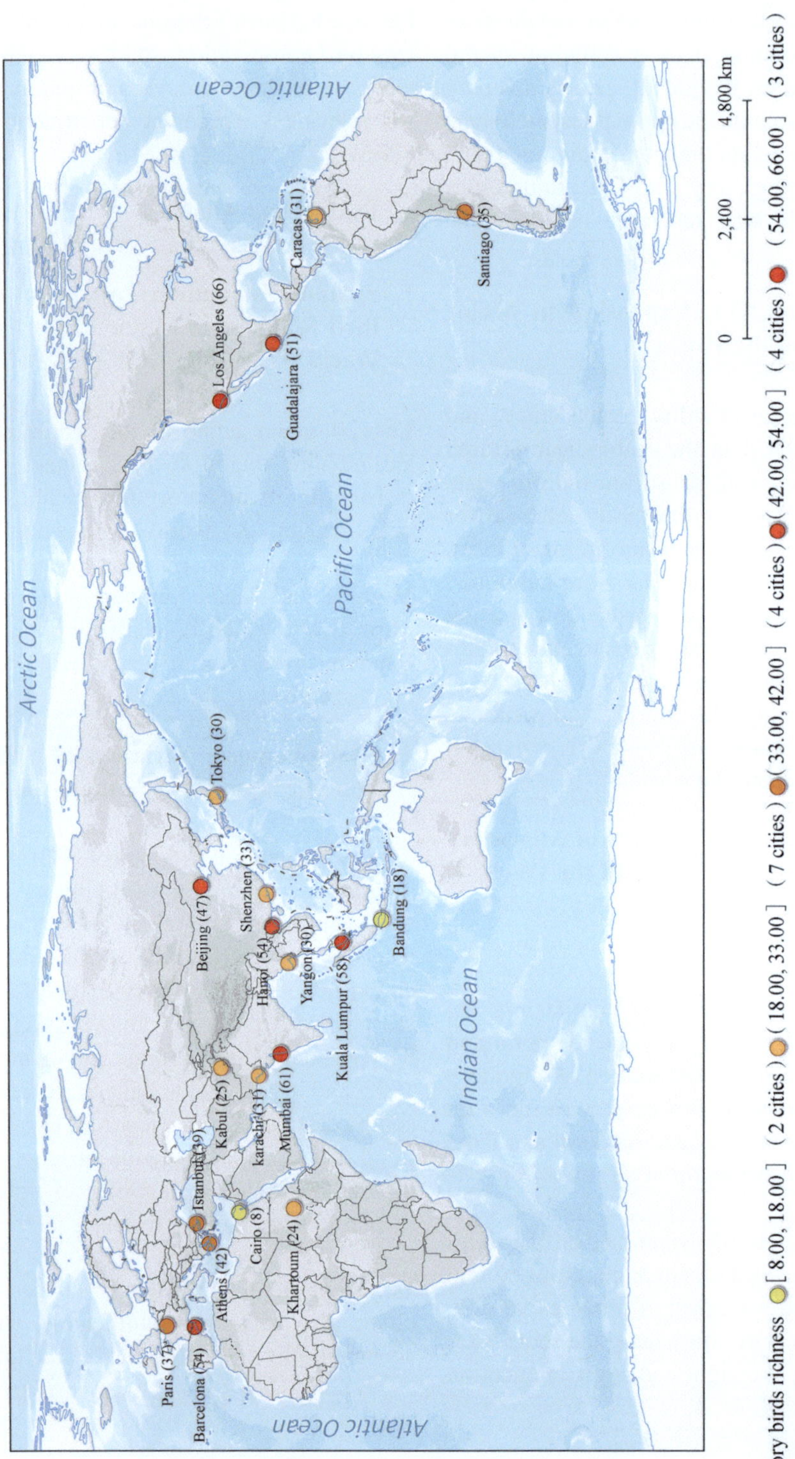

Fig. 8.7 Spatial distribution map of migratory bird richness in major cities around the world

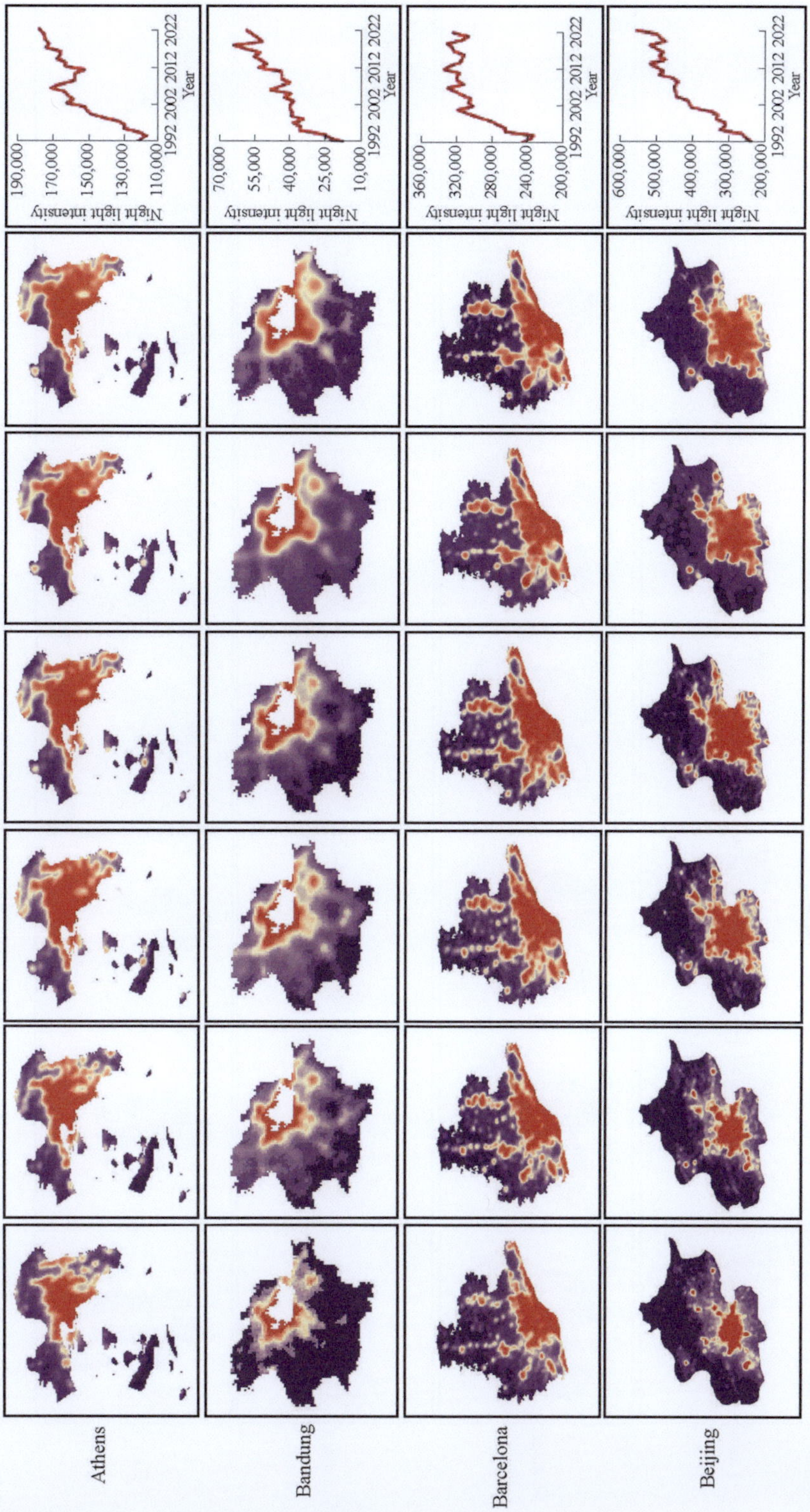

Fig. 8.8 Spatiotemporal variation of NTL expansion in major cities around the world from 1992 to 2022

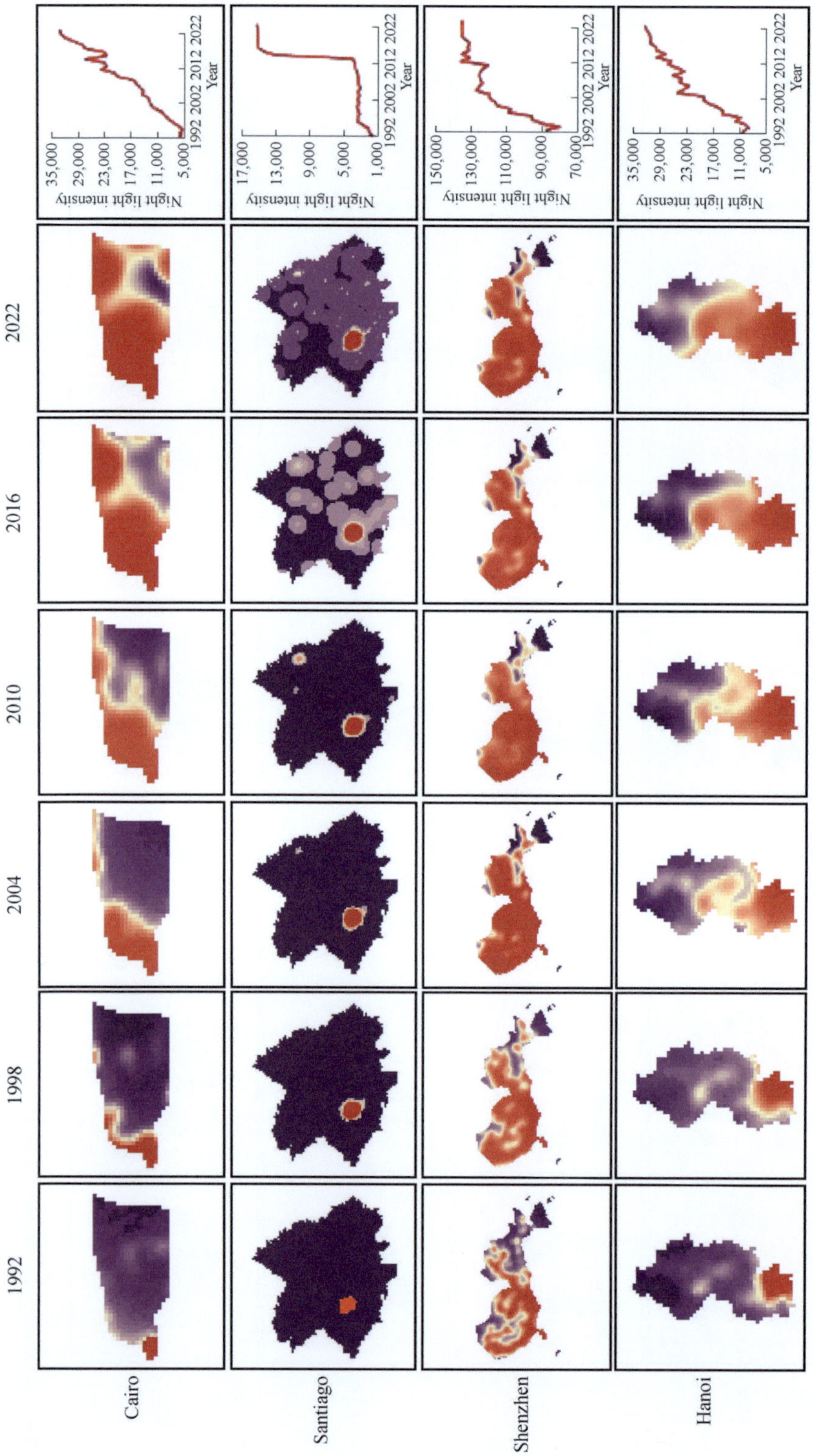

Fig. 8.8 (continued)

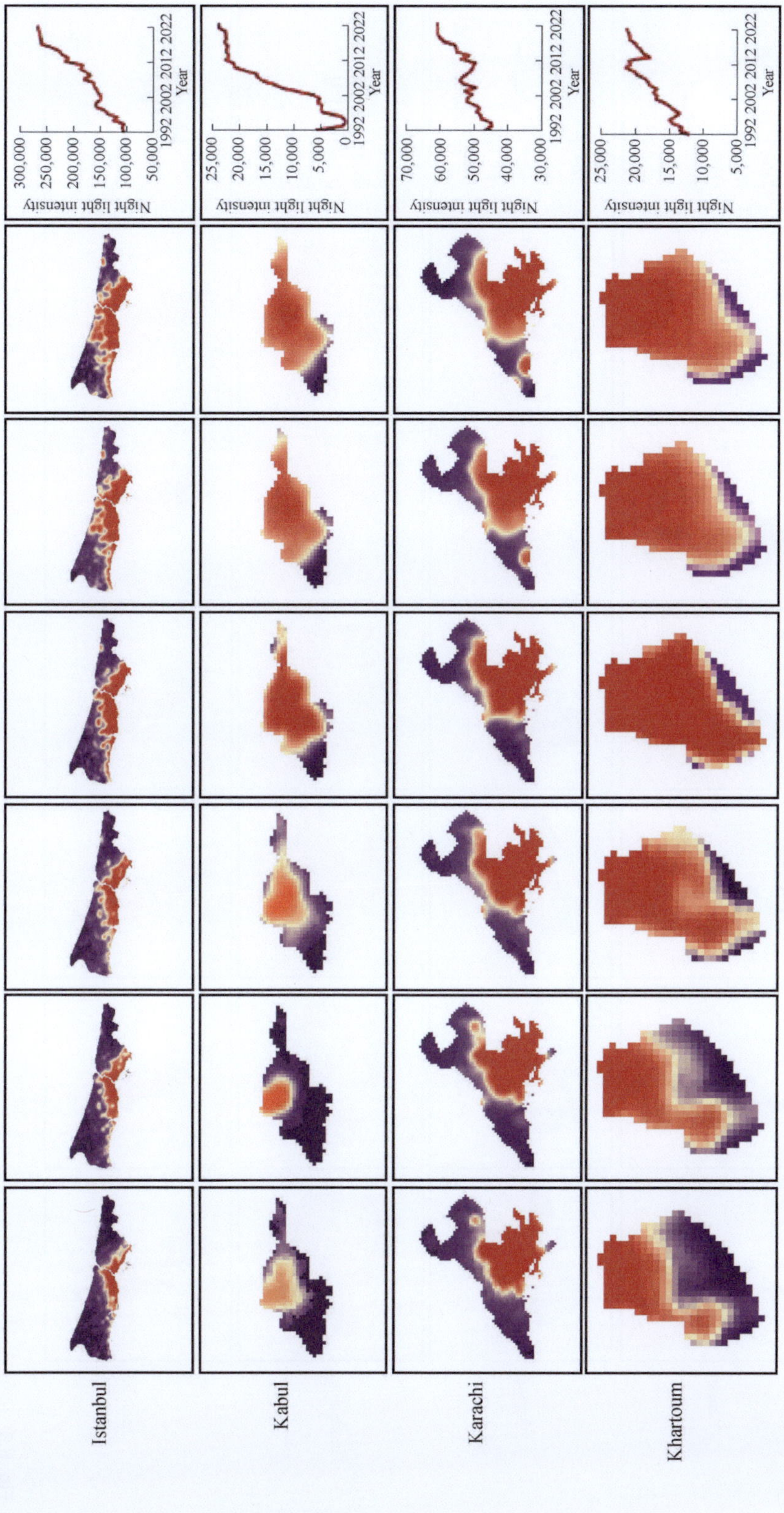

Fig. 8.8 (continued)

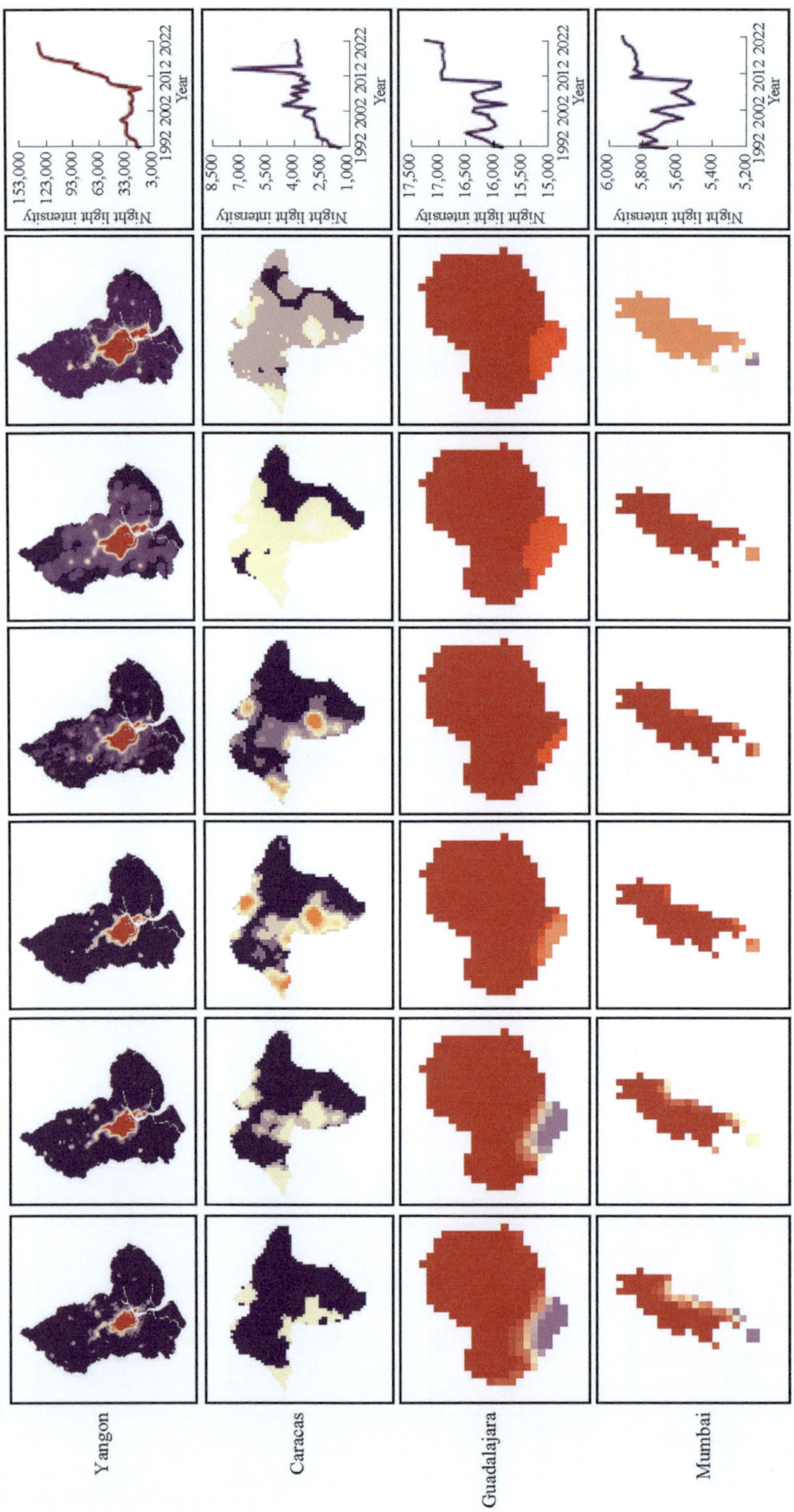

Fig. 8.8 (continued)

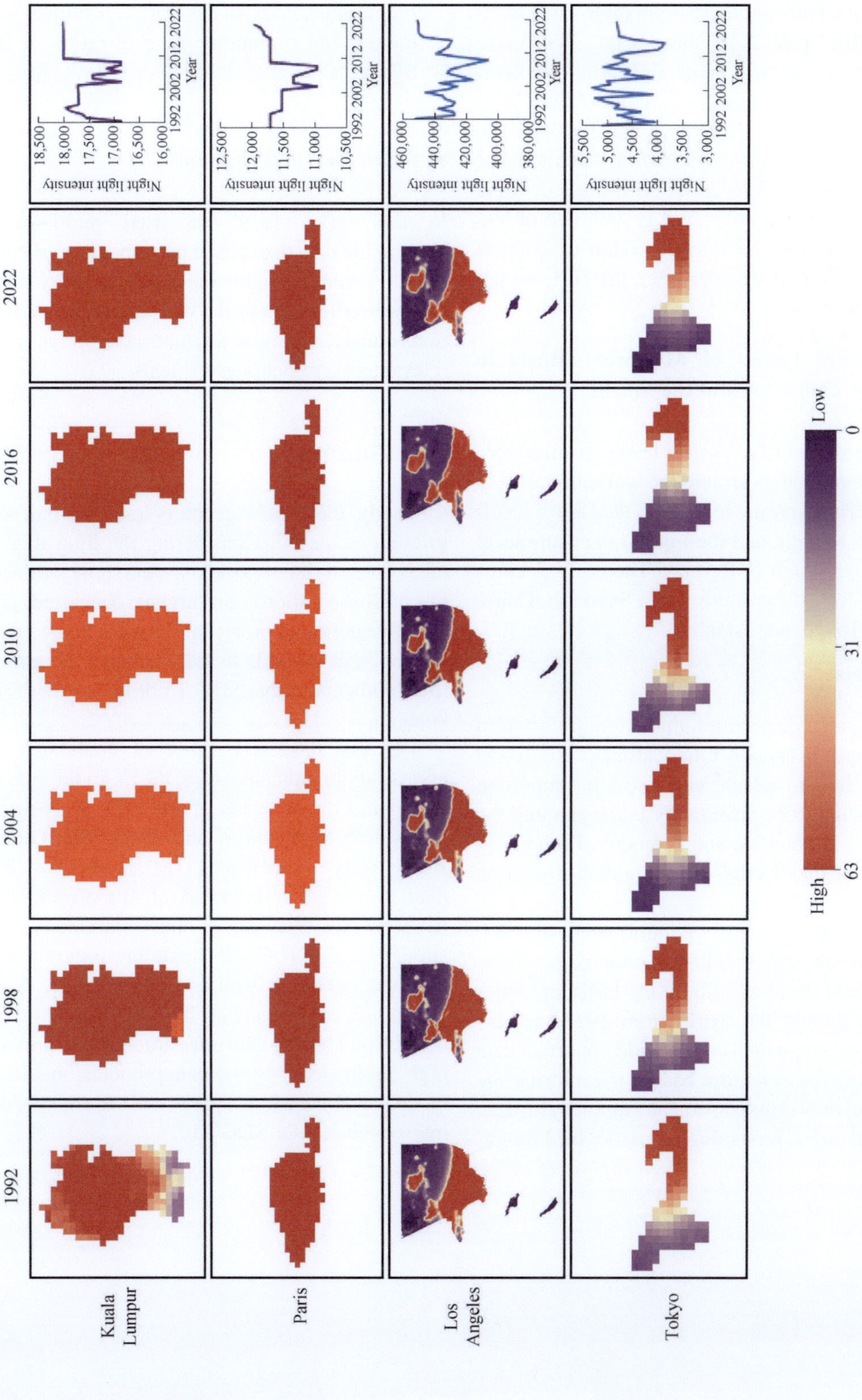

Fig. 8.8 (continued)

2022. It should be noted that Yangon, Myanmar, has seen a significant increase in high-value areas since 2010. Since 2016, San Diego's NTL have been high in sporadic areas, indicating acceleration in the expansion of NTL in all cities. Cities such as Guadalajara, Karachi, Kuala Lumpur, and Mumbai saw only a small expansion of nighttime lighting from 1992 to 2022, with most of these cities experiencing an overall growth rate of less than 10%. In some developed countries, such as Tokyo, Paris, Los Angeles, etc., the NTL do not change much.

3. **Exposure Levels of Migratory Birds in Major Cities Around the World**

The distribution of exposure levels of migratory birds in major cities around the world is shown in Fig. 8.9. The exposure level is divided into 5 levels from low to high, and the highest exposure level has 1 city, which is Beijing. There are 5 cities with the lowest exposure level: Santiago, Cairo, Athens, Tokyo, and Istanbul.

Highlights

- The spatial distribution map of migratory bird richness in major cities around the world was drawn, which can provide important data support for accurately understanding the spatial distribution and diversity of migratory birds in major cities on the global migration routes.
- The spatiotemporal variation map of NTL expansion and the distribution map of the exposure level of migratory birds in major cities around the world from 1992 to 2022 were drawn, which can provide important data support and scientific basis for exploring the impact of NTL expansion on migratory birds in global cities, protecting migratory bird migration to maintain global bird diversity and harmonious coexistence between human and nature, and promoting the realization of the SDGs.

- **Discussion and Outlook**

In this case study, we used multi-source geographic data to evaluate the expansion of NTL and the exposure assessment of migratory birds in major cities around the world, and the evaluation results can provide an important reference for relevant government departments.

8.4 Summary

Currently, the 2030 Agenda is less than halfway to being achieved. Considering the long time it takes for ecosystems to recover and show the benefits of conservation, only urgent, concerted, and effective action to protect and restore ecosystems will make it possible to fully achieve the terrestrial biodiversity that SDG 15 brings to the 2030 Agenda.

We still have a number of challenges in assessing the indicators of SDG 15. In the future, facing the major needs of the "UN Decade on Ecosystem Restoration" and "Post-2020 Global Biodiversity Framework", we should give full play to the great potential of Big Earth Data in SDG 15 monitoring and assessment, and strengthen the SDGSAT-1 satellite program and its applications, as well as other satellites from China and accompanying big data platforms for the SDGs. This big data infrastructure can provide high-quality, easy-to-use data products, methods, models, and decision support tools to support the implementation of SDG 15.

Fig. 8.9 Exposure levels of migratory birds in major cities around the world

References

FAO, UNEP (2020) The state of the world's forests 2020: forests, biodiversity and people [M]. Rome: FAO

IPBES (2019) Global assessment report on biodiversity and ecosystem services of the intergovernmental science-policy platform on biodiversity and ecosystem services [R]. https://doi.org/10.5281/zenodo.3831673 [2023-05-30]

United Nations Environment Programme (2021a) Measuring progress: environment and the SDGs [R]. https://wedocs.unep.org/bitstream/handle/20.500.11822/35968/SDGMP.pdf [2023-07-08]

United Nations (2019) The sustainable development goals report 2019 [R]. https://www.undp.org/sites/g/files/zskgke326/files/migration/ao/55b904f4aa8acbed50805210654424d983c917af82b941d7f2ab4e007043995f.pdf [2024-10-09]

Yang XD (2020) Overview of desertification and desertification land monitoring in inner mongolia autonomous region [J]. Inn Mongla For Investig Des 43(2):86–88 (In Chinese)

Yuan WP, Liu SG, Yu GR et al (2010) Global estimates of evapotranspiration and gross primary production based on MODIS and global meteorology data[J]. Remote Sens Environ 114(7):1416–1431

9 Integrated Evaluations and Interactions Among SDGs

9.1 Background

Globally, the achievement of the SDGs faces numerous challenges, one of which is how to effectively monitor and assess progress. The UN's 2030 Agenda encompasses a broad range of goals, from eradicating extreme poverty to protecting the natural environment. A key to achieving these goals lies in the use of innovative tools and methods, with Big Earth Data being an indispensable resource.

Big Earth Data consists of large datasets associated with spatial attributes and Earth science, including data on land, oceans, atmosphere, and human activities. It is characterized by massive volume, multi-source, and multi-temporal dimensions (Guo et al. 2021; Dong et al. 2021), offering significant advantages in interaction and integrated studies of the SDGs at the urban scale. On the one hand, Earth observation and geographic information data provide continuous spatial and temporal coverage, addressing inconsistencies in the standards and quality of statistical data at the urban scale. These data can promptly capture changes in urban surface elements, and their high spatial resolution offers more detailed and precise urban surface information, providing critical insights for SDG monitoring at the urban scale. On the other hand, geographic information modeling and simulation methods, based on spatial analysis and other techniques, can effectively uncover the interactions between different urban SDGs, revealing their potential mechanisms and key factors (Xiao et al. 2022). Furthermore, constructing predictive models for future sustainable development that account for inter-city SDG interactions, and dynamically simulating the achievement of the SDGs under different future development scenarios, can provide scientific support for future urban planning and decision-making.

The first case study in this chapter constructs a system dynamics model to analyze the future development scenarios under the interaction of multiple SDG indicators globally, predicting the progress of SDG achievement in various scenarios by 2030. Using data from 148 countries globally (2000–2020) and scenario data from the Commentary on the Coupled Model Intercomparison Project Phase 6 (CMIP6), the case study examines the performance of six key indicators—SDG 1.1.1, SDG 2.1.1, SDG 6.1.1, SDG 7.1.1, SDG 11.6.2, and SDG 13.2.2—under different development scenarios. The second case study utilizes Big Earth Data to explore the trade-offs between ecosystem services and human activities in waterfront areas at both global and Chinese scales. Based on relevant SDG indicators, the case study provides comprehensive insights into the sustainability of waterfront areas and offers predictions for future development trends. The third case study analyzes global economic inequality, particularly in the region involved in the BRI, using NTL remote sensing imagery, airflow data, and other Big Earth Data. The case study employs the Theil

H. Guo, *Big Earth Data in Support of the Sustainable Development Goals (2023)—The Belt and Road*, Sustainable Development Goals Series, https://doi.org/10.1007/978-981-95-3178-3_9

index decomposition method to conduct a multi-stage analysis of economic disparity patterns from 2000 to 2020. It reveals trends in economic inequality both globally and within the region involved in the BRI, including inter-country and intra-country disparities, while highlighting the role of the BRI in reducing economic inequality.

9.2 Main Contributions

To clarify the complex synergies and trade-offs among SDGs, this chapter primarily focuses on the integrated evaluations and interactions among SDGs on the global scale, simulation and prediction of green and coordinated development of global waterfront areas, and spatiotemporal analysis of economic development equilibrium. It showcases the technical advantages of Big Earth Data in supporting SDG integration and interactions on a global scale. This chapter contributes mainly in two areas: methods and models, and decision support (Table 9.1).

9.3 Case Study

9.3.1 Future Development Scenarios Forecast Under the Multi-indicator Interactions of Global SDGs

Targets: SDG 1.1.1, 2.1.1, 6.1.1, 7.1.1, 11.6.2, 13.2.2.

- **Background**

The SDGs are a comprehensive framework of human, social, and environmental development

Table 9.1 Cases and their main contributions

Targets	Cases	Contributions
SDG 1.1.1, 2.1.1, 6.1.1, 7.1.1, 11.6.2, 13.2.2	Future Development Scenarios Forecast under the Multi-Indicator Interactions of Global SDGs	**Method and model**: Using a system dynamics approach, we developed a predictive model for the interactions among SDG indicators. By integrating the CMIP6 SSP-RCP scenarios, we forecast the sustainable realization process and development trend for SDG 1.1.1, SDG 2.1.1, SDG 6.1.1, SDG 7.1.1, SDG 11.6.2, and SDG 13.2.2 under various scenarios by 2030 **Decision support**: Some developing countries still need to accelerate economic growth, actively eradicate poverty, achieve Zero Hunger, and enhance water resource management. Meanwhile, some upper-middle-income countries need to adopt low-carbon and emission-reducing consumption and production patterns to promote environmental sustainability
SDG 6.6, 8.9, 11.3, 15.1	Simulation and Prediction of Green and Coordinated Development of Global Waterfront Areas	**Method and model**: This case study utilizes an integrated approach combining the Integrated Valuation of Ecosystem Services and Trade-offs (InVEST model) for ecosystem services, RStudio modeling, the PLUS model for land-use change, and GIS spatial analysis techniques to explore the trade-offs and synergies between ecosystem services and human activities in global waterfront areas from 2010 to 2020 **Decision support**: Waterfront areas in regions such as southern Beijing, Tianjin, northeastern Hebei, northwestern Shandong, northeastern Henan, western Jiangsu, northern Ningxia, and western Xinjiang in China require enhanced ecological restoration and a reduction in the intensity of human activities. These actions are crucial for promoting sustainable development and the construction of riparian tourism in these areas
SDG 10	Spatiotemporal Analysis of Economic Development Equilibrium Based on Big Earth Data	**Method and model**: NTL remote sensing imagery data were used to generate per capita economic development data at various spatial scales globally. These data serve as a basis for assessing the trends in global economic inequality from 2000 to 2020 **Decision support**: The BRI has contributed a Chinese solution to addressing regional imbalances, accounting for over 70% of the reduction in such inequalities. Before the BRI was introduced, global economic inequality decreased at an average annual rate of 1.4%. Following the launch of the BRI, the annual rate of reduction accelerated to 2.1%, representing an improvement of more than 50%

goals, which include 17 goals, 169 targets, and more than 230 indicators. The prediction of the development trend of the SDGs under different scenarios is crucial for the current resource allocation, policy optimization and promotion of the 2030 Agenda in the region (Moyer and Hedden 2020). At present, the predictive modeling of the SDGs has become a hot topic of research. The Bill & Melinda Gates Foundation's Goalkeepers Report used historical data and economic growth rates to predict the progress of the 2030 goals at the global level, and the Millennium Institute used the system dynamics approach to construct the iSDG model, which can reflect the complex interactions and feedback mechanisms among multiple indicators of the SDGs (Allen et al. 2019), in order to make SDGs future projections.

The case study constructed the system dynamics model that takes into account the interactions among multiple indicators of the SDGs, integrates different SSP-RCP combination scenarios in CMIP6, and predicted the sustainable realization process and development trend of SDG 1.1.1, SDG 2.1.1, SDG 6.1.1, SDG 7.1.1, SDG 11.6.2, and SDG 13.2.2 indicators in 2030 under the five SSP-RCP combination scenarios globally. This will provide decision support for future development.

- **Data**
  - Statistical data on SDG indicators for 148 countries obtained by the UN, 2000–2020 (https://unstats.un.org/sdgs/dataportal/database/).
  - Statistical data on SDG indicators for 148 countries obtained by the WB, 2000–2020 (https://databank.worldbank.org/source/world-development-indicators/).
  - Global cropland dataset, 2020–2030 (https://zenodo.org/records/5105689).
  - Global temperature and precipitation CMIP6 scenario data, 2020–2030 (http://esgf-node.llnl.gov/search/cmip6/).

- **Method**

The system dynamics model was developed to account for the complex interactions and feedback mechanisms among SDG indicators. This model was constructed using historical data from 148 countries during 2000–2020, and it involves 45 variables (19 SDG indicator variables and 26 non-SDG indicator variables) in social, economic, and environmental aspects. This model integrated various SSP-RCP combination scenarios from CMIP6, choosing five SSP-RCP scenarios (SSP1-RCP2.6, SSP2-RCP4.5, SSP3-RCP7.0, SSP4-RCP6.0, and SSP5-RCP8.5) to predict the future development process of the six SDG indicators (SDG 1.1.1, SDG 2.1.1, SDG 6.1.1, SDG 7.1.1, SDG 11.6.2, and SDG 13.2.2).

The realization process and development trend of the SDG indicators under different scenarios for each country in the world in 2030 were assessed based on the SDGs indicator panels in the *Sustainable Development Goals Report 2022* (United Nations 2022). The analysis encompassed the examination of scenario-related variations and spatial differences in the realization process of SDG indicators. The case study identified the optimal SSP-RCP development scenarios for various indicators within each nation, with the objective of providing informed guidance for the implementation of the global SDGs.

- **Results and Analysis**

Figure 9.1 illustrates the process of achieving the SDGs and the development trends in typical countries under different scenarios. The majority of the countries have demonstrated substantial progress in SDG 1.1.1 (proportion of the population living below the international poverty line), SDG 2.1.1 (prevalence of undernourishment), SDG 6.1.1 (proportion of population using safely managed drinking water services), and SDG 7.1.1 (proportion of population with access to electricity), with the objective of

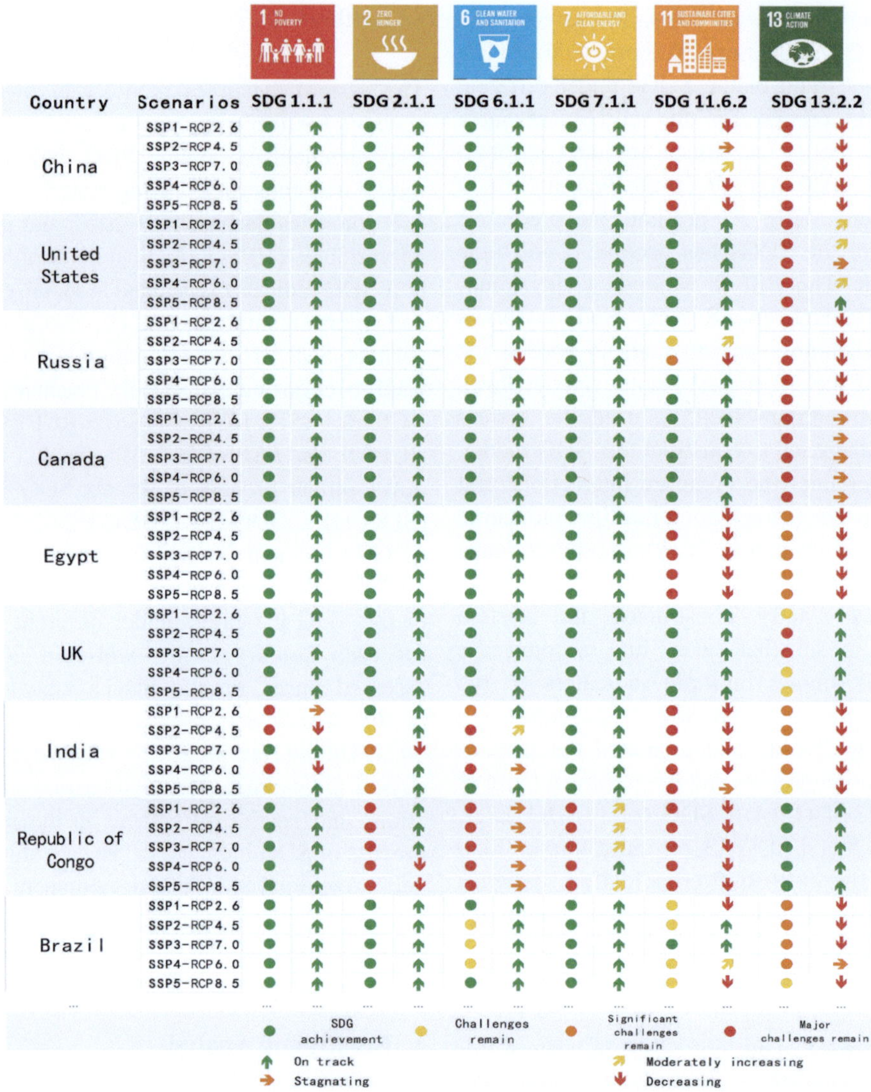

Fig. 9.1 SDG realization process and development trends in typical countries under different scenarios

achieving sustainable targets. However, SDG 11.6.2 (annual mean levels of fine particulate matter, e.g., PM2.5) and SDG 13.2.2 (total greenhouse gas emissions per year) have exhibited slow progress or even stagnation, encountering varying degrees of challenges.

The realization of SDG 1.1.1, SDG 2.1.1, and SDG 6.1.1 in some developing countries is still hindered by numerous challenges, and significant progress is still required in the areas of poverty alleviation, achieving Zero Hunger, and strengthening water resource management. Furthermore, there is an urgent need to accelerate economic development and promote the realization of SDG indicators in the category of basic needs in developing countries. While some upper-middle-income countries have largely achieved the SDGs outlined in SDG 11.6.2, significant challenges persist in the realization of SDG 13.2.2. These countries must adopt low-carbon and emission-reducing consumption and production patterns, and promote environmentally sustainable development.

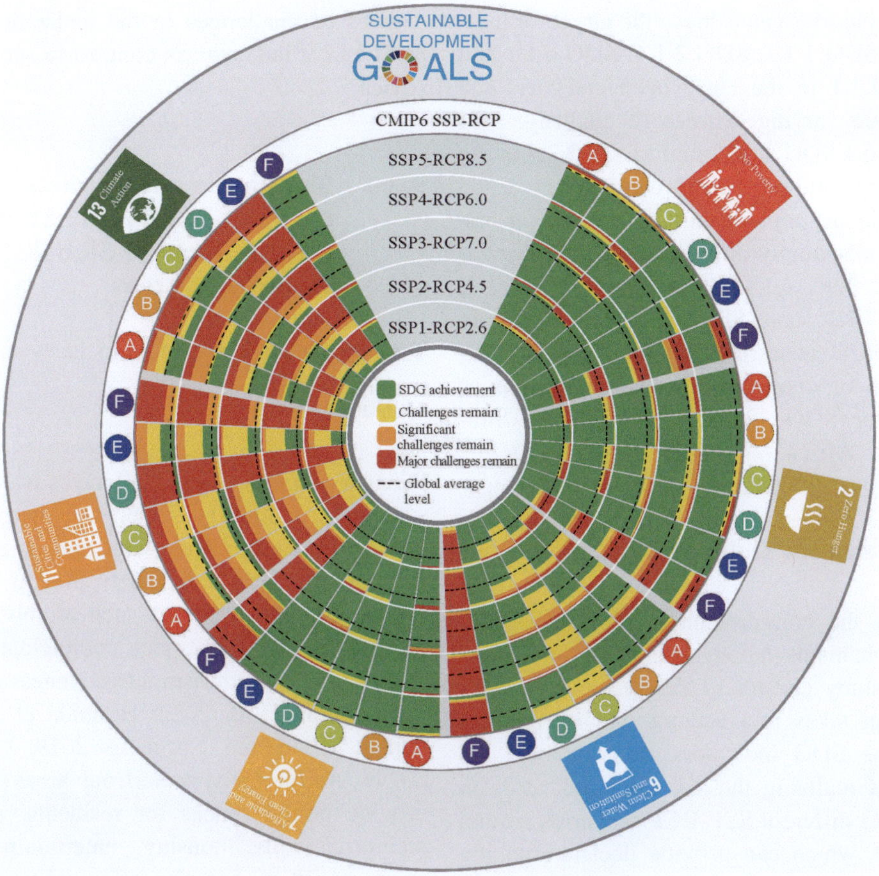

Fig. 9.2 Projection results of the SDG indicators for the six global regions under different scenarios. *Note* A to F are East and South Asia, Eastern Europe and Central Asia, Latin America and the Caribbean, the Middle East and North Africa, the countries of the OECD, and Sub-Saharan Africa sequentially

The projected results of the SDG indicators for the six global regions under different scenarios are shown in Fig. 9.2. By 2030, the vast majority of countries in the world can achieve the SDGs in the basic need category of SDG 1.1.1, SDG 2.1.1, and SDG 7.1.1; about 50% of the countries can achieve SDG 6.1.1; however, most of the countries in the environmental protection category of SDG 11.6.2 and SDG 13.2.2 face varying degrees of challenges. In addition, there are spatial differences in the process of achieving the global SDGs. Among the six projected SDG indicators, sub-Saharan Africa is in a better position to achieve SDG 13.2.2, but there are varying degrees of challenges in achieving SDG 1.1.1, SDG 2.1.1, SDG 6.1.1, SDG 7.1.1, and SDG 11.6.2, in contrast to the countries of the Organization for Economic Cooperation and Development (OECD).

Highlights

- The system dynamics approach was used to construct an interaction prediction model for the SDG indicators. Combining CMIP6 SSP-RCP scenarios, this case study predicts the sustainable development progress of SDG 1.1.1, SDG 2.1.1, SDG 6.1.1, SDG 7.1.1, SDG 11.6.2, and SDG 13.2.2 indicators under five scenarios for 2030. It further compares the scenario-based and spatial differences in achieving these indicators, providing decision-making support for better global realization of the SDGs.

- Most countries can achieve the target for indicators SDG 1.1.1, SDG 2.1.1, SDG 6.1.1 and SDG 7.1.1 in the basic need category, and there are varying degrees of challenges for indicators SDG 11.6.2 and SDG 13.2.2 in the environmental protection category.
- The global implementation process of the SDGs is spatially differentiated. The realization of SDG 1.1.1, SDG 2.1.1, SDG 6.1.1, SDG 7.1.1, and SDG 11.6.2 is better in the OECD countries, but there are varying degrees of challenges in the realization of indicator SDG 13.2.2, in contrast to Sub-Saharan Africa.

- **Discussion and Outlook**

Based on the complex interactions and feedback mechanisms among the SDG indicators, the case study constructd the system dynamics model that takes into account the interactions among the SDG indicators, and explores the process of realizing the six SDG indicators in 2030 under different SSP-RCP scenarios around the world, which can provide decision-making support for realizing the global SDGs. The case study found that: (1) Most countries can achieve SDG 1.1.1, SDG 2.1.1, SDG 6.1.1, and SDG 7.1.1 in the basic need category, and there are varying degrees of challenges for indicators SDG 11.6.2 and SDG 13.2.2 in the environmental protection category. (2) Some developing countries still need to accelerate economic development, actively eradicate poverty, achieve Zero Hunger and strengthen water resource management to contribute to the development of the three basic need categories of SDG 1.1.1, SDG 2.1.1 and SDG 6.1.1. Some upper-middle-income countries need to adopt low-carbon and emission-reducing consumption and production patterns to promote environmentally sustainable development and accelerate the process of realizing SDG 13.2.2. (3) There are spatial differences in the realization process of global SDGs. The realization of SDG 1.1.1, SDG 2.1.1, SDG 6.1.1, SDG 7.1.1, and SDG 11.6.2 is better in the OECD countries, but there are varying degrees of challenges in the realization of the SDG 13.2.2 indicator, in contrast to sub-Saharan Africa.

9.3.2 Simulation and Prediction of Green and Coordinated Development of Global Waterfront Areas

Targets: SDG 6.6, 8.9, 11.3, 15.1.

- **Background**

Waterfront areas, where land meets water, represent unique interfaces between humans and the environment. They are hybrid zones of ecology, economy, and society, playing a crucial role in economic aspects such as job creation, urban vitality enhancement, commercial activity stimulation, and tourism attractiveness enhancement (Ferah et al. 2022; Hermida et al. 2019; Papatheochari and Coccossis 2019; Qu et al. 2024). Additionally, waterfront areas not only provide public spaces for residents, including memorial halls, housing, entertainment, and sports facilities, but also offer political, social, environmental, and cultural values. These regions serve as habitats for wildlife and contribute to mitigating the urban heat island effect (Avni and Fischler 2020).

Therefore, the regulation of global waterfront areas is crucial. This case study adopted an interdisciplinary perspective incorporating four indicators—SDG 6.6, SDG 8.9, SDG 11.3, and SDG 15.1—and focused on global waterfront areas to explore the evolution and trade-offs and synergies between four ecosystem service functions (annual water yield, carbon storage and sequestration, sediment delivery ratio, and habitat quality) and human activities from 2010 to 2020. Furthermore, this case study delved into the relationship between ecosystem service functions and human activities under different land use scenarios in China's waterfront areas, providing decision-making support for the sustainable development of global waterfront areas and the high-quality development of China's waterfront areas. This, in turn, promotes sustainable tourism development and sustainable urban construction in waterfront areas.

- **Data**
  - Global DEM data in 2010, 2015, and 2020, with a spatial resolution of 90 m (https://www.sciencedirect.com/science/article/pii/S2095927322005412?via%3Dihub); global land cover/land use data, with a spatial resolution of 300 m (https://cds.climate.copernicus.eu/cdsapp#!/dataset/satellite-land-cover?tab=overview); rainfall data, with a spatial resolution of 1 km (https://www.nature.com/articles/s41597-020-0453-3); actual evapotranspiration data at the surface, with a spatial resolution of 1 km (https://poles.tpdc.ac.cn/en/data/0c9ce00c-6ee4-4ee0-a99c-67a833975 8ea/?q =); world soil database data, with a spatial resolution of 1 km (https://www.fao.org/soils-portal/data-hub/soil-maps-and-databases/harmonized-world-soil-database-v12/en/).
  - Spatial resolution data on vegetation coverage in China in 2010, 2015, and 2020, with a spatial resolution of 30 m (http://www.gis5g.com/data/zbsj/FVC?id=20); NTL data, with a spatial resolution of 500 m (https://eogdata.mines.edu/products/vnl/); population density data, with a spatial resolution of 1 km (https://hub.worldpop.org/ project/categories?id = 18); road data (https://www.webmap.cn/commres.do?method = result25W).
  - Global GHS Settlement Model grid (GHS-SMOD) data in 2010, 2015, 2020, 2025, and 2030, which represents high-precision population density and building area data, with a spatial resolution of 1 km (https://human-settlement.emergency.copernicus.eu/ download.php?ds = smod); GHS Built-Up Volume Grid (GHS-BUILT-V) data, which represents high-precision building volume data, with a spatial resolution of 1 km (https://human-settlement.emergency.copernicus.eu/download.php?ds=builtV).

- **Method**

This case study utilized ArcGIS 10.6 and the InVEST model, developed by Stanford University, to calculate and visualize four ecosystem service functions in global waterfront areas for the years 2010, 2015, and 2020. These services include annual water yield, carbon storage and sequestration, sediment delivery ratio, and habitat quality. By overlaying these four ecosystem services with equal weights, the total global ecosystem service supply was computed. Subsequently, the natural break method was used to classify the total ecosystem service supply, and changes from 2010 to 2020 were analyzed. The RStudio model was employed to explore the trade-offs and synergies between ecosystem services and human activiies (GHS-SMOD, GHS-BUILT-V) in global waterfront areas. Using the PLUS model, land-use spatial patterns in China's waterfront areas for 2025 and 2030 were simulated under three scenarios: natural scenario, ecological conservation scenario, and integration of culture and tourism scenario. Based on this, the case study examined the impact of land-use changes on ecosystem services under different scenarios and analyzed the relationship between these changes and human activity intensity.

- **Results and Analysis**

1. **Spatiotemporal Changes and Trade-Offs and Synergies of Ecosystem Services in Global Waterfront Areas from 2010 to 2020**

The changes in ecosystem services in global waterfront areas from 2010 to 2020 are illustrated in Fig. 9.3. In terms of annual water yield, there was a trend of initial decrease followed by an increase, with an average annual value decreasing from 125 mm in 2010 to 111 mm in 2015 and then increasing to 128 mm in 2020. Regarding carbon storage and sequestration, there was a downward trend from 2010 to 2020, decreasing from 128.8 billion tons in 2010 to 127.8 billion tons in 2020. Meanwhile, sediment

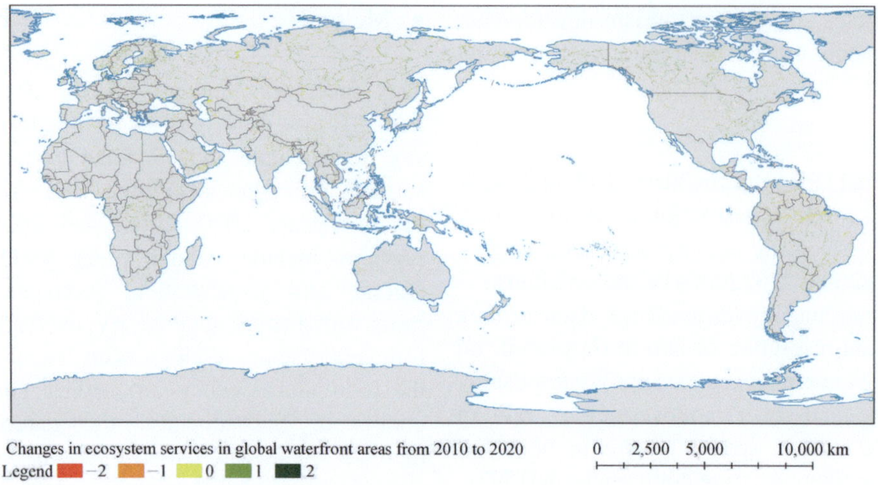

Fig. 9.3 Changes in ecosystem services in global waterfront areas from 2010 to 2020

delivery ratio increased from 2.6 billion tons in 2010 to 2.7 billion tons in 2020. In terms of habitat quality, it remained basically unchanged from 2010 to 2020, with an average score of approximately 0.81. Countries and regions with significant growth trends in ecosystem services mainly include Norway, Sweden, Gabon, the Democratic Republic of the Congo, eastern Russia, central and southern China, northeastern India, eastern Myanmar, northeastern Cambodia, Canada, northwestern and southeastern United States, eastern Peru, northeastern Bolivia, and northern Brazil (Fig. 9.3).

From 2010 to 2020, global waterfront areas exhibited significant negative correlations between (1) annual water yield and habitat quality, (2) carbon storage and sequestration versus building volume, and (3) habitat quality versus population density-builting area, demonstrating pronounced trade-off relationships among these variables. During this period, the trade-off between carbon storage and sequestration and building volume in global waterfront areas intensified, while the trade-offs between habitat quality and building volume, as well as habitat quality and population density-building area, weakened (Fig. 9.4).

The relationship between ecosystem services and building volume in global waterfront areas is complex. From 2010 to 2015, there was a trade-off relationship, which turned into a synergistic relationship from 2015 to 2020. Furthermore, the synergistic effect between ecosystem services and population density-building area gradually increased from 2010 to 2020. The main reason for this may be that within ecosystem services, high-value areas for annual water yield and soil retention are mainly concentrated in construction land areas, while high-value areas for carbon storage and sequestration and habitat quality are mainly concentrated in regions with high vegetation cover such as forests and grasslands. From 2010 to 2015, there was a significant downward trend in annual water yield, and carbon storage and sequestration, resulting in a gradual reduction in trade-offs. However, from 2015 to 2020, annual water yield and soil retention increased significantly, while carbon storage and sequestration decreased, indicating an increased impact of construction land. Therefore, the synergistic effect of building volume and population density-building area gradually increased.

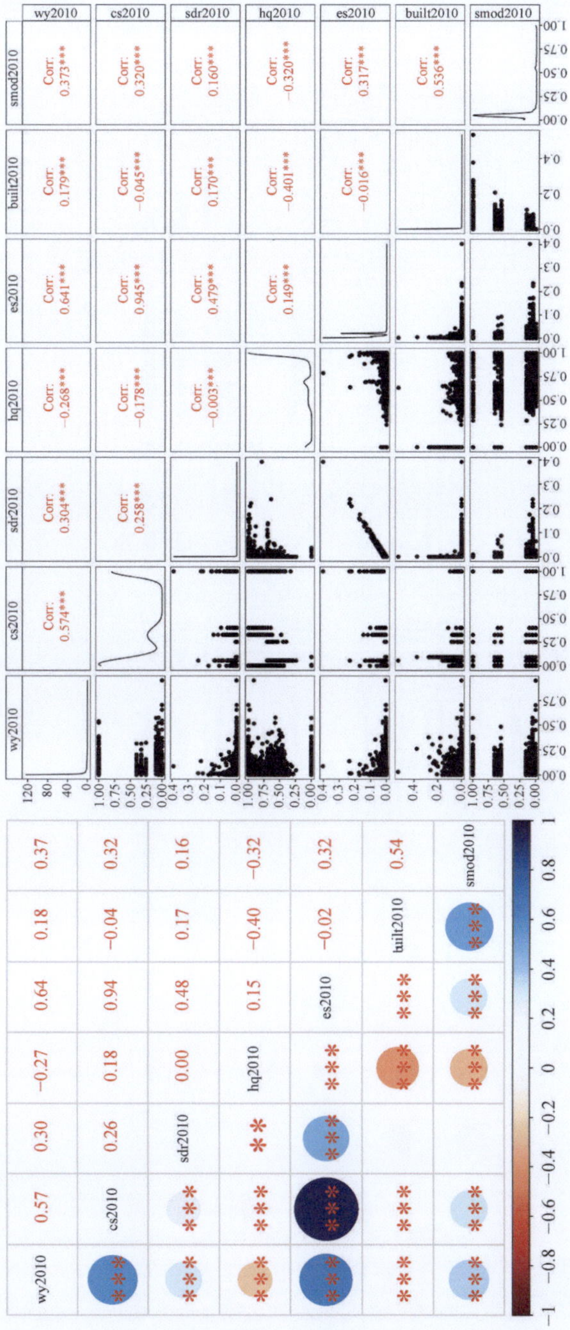

(a) 2010

Fig. 9.4 Diagram of trade-offs and synergies between ecosystem services and human activities in global waterfront areas. *Note* wy: annual water yield;, cs:carbon storage and sequestration; sdr: sediment delivery ratio;hq:habitat quality; es:ecosystem service; built: GHS-BUILT-V; smod: GHS-SMOD

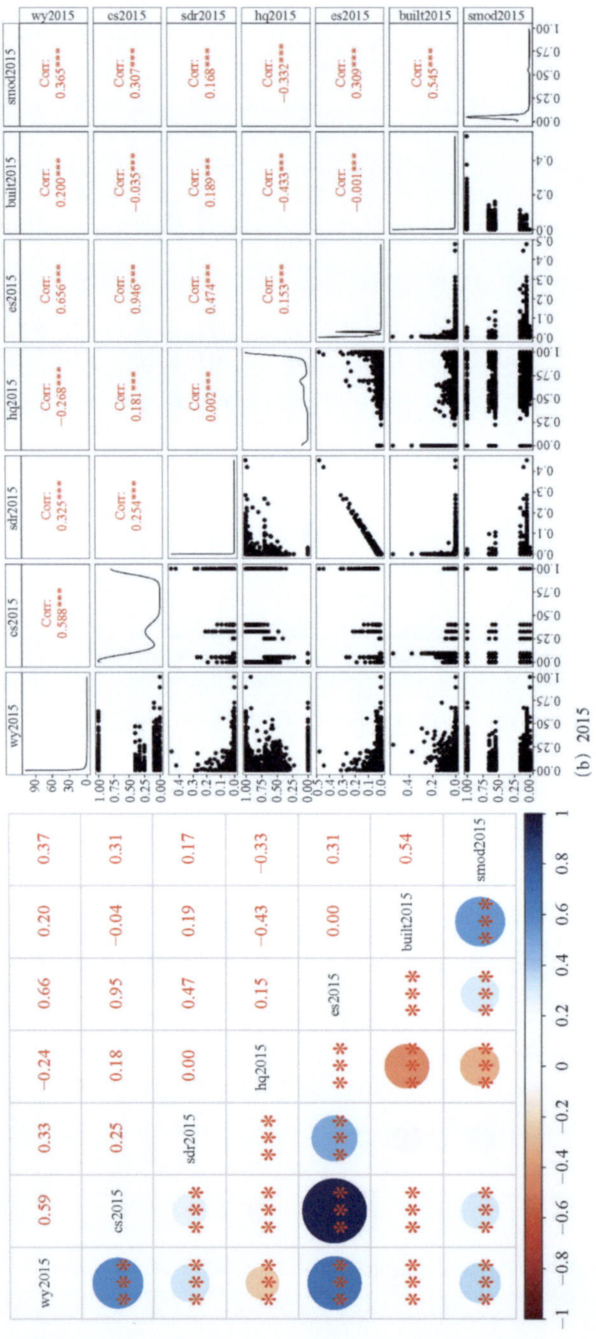

Fig. 9.4 (continued)

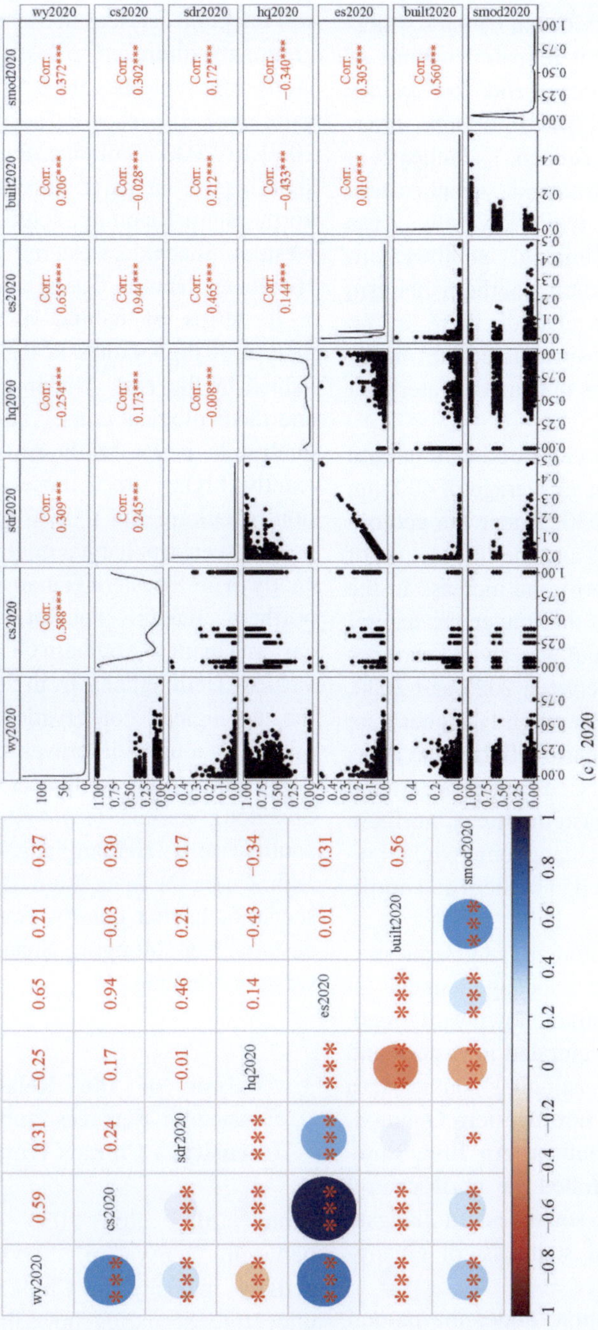

(c) 2020

Fig. 9.4 (continued)

2. **Changes in Ecosystem Services Under Different Land Use Scenarios in China's Waterfront Areas from 2025 to 2030**

The overall ecosystem services in China's waterfront areas in 2030 exhibit a spatial pattern of high values in the southeast and low values in the northwest (Fig. 9.5). High-value areas are concentrated in Yunnan, southeastern Xizang, northwestern Guangxi, southeastern Jilin, and western Hubei, while low-value areas are concentrated in Xinjiang, southwestern Qinghai, western Nei Mongol, southern Jiangsu, and southern Guangdong. Overall, there are no significant changes in ecosystem services under different land use scenarios in China's waterfront areas from 2025 to 2030.

In terms of annual water yield, under the natural scenario, it decreases from an average of 400 mm in 2020 to 396 mm in 2030. Under the ecological conservation scenario and the integration of culture and tourism scenario, the increase in the forest area is greater than that under the natural scenario. As a result, annual water yield services remain relatively stable between 2025 and 2030, averaging 397 mm annually, slightly higher than that under the natural scenario. High-value areas for annual water yield services are concentrated in southern Hubei, northeastern Hunan, northern Jiangxi, and southeastern Guangdong, while low-value areas are concentrated in western and northeastern China.

In terms of carbon storage and sequestration, it remains basically unchanged under all three scenarios, with a slight downward trend. High-value areas are concentrated in eastern and southern China, mainly including southeastern Xizang, western Yunnan, northwestern Guangxi, western Zhejiang, and southeastern Jilin. Low-value areas are concentrated in north-central Xizang, eastern Qinghai, southwestern Jiangsu, northeastern Jiangxi, and southeastern Guangdong.

In terms of soil retention, under the natural scenario, there is no significant change in soil retention services, with high-value areas mainly concentrated in southeastern Xizang, northwestern Yunnan, and southwestern Taiwan. The spatial distribution under the ecological conservation scenario is similar to that under the natural scenario, but there is a downward trend from 2020 to 2030, with a significant increase in soil retention services in Sichuan. Under the integration of culture and tourism scenario, the impervious area increases by 0.53% compared to the natural scenario, showing an increasing trend from 2020 to 2030. Soil retention services improve significantly in parts of southwestern Beijing, northeastern Liaoning, southern Qinghai, southwestern Shaanxi, western Hubei, southeastern Hunan, and central Guangxi.

In terms of habitat quality, it exhibits a pattern of high values in the southwest and low values in the east. Under the natural scenario and the ecological conservation scenario, habitat quality is better, with grassland areas significantly higher than those under the integration of culture and tourism scenario. Under the natural scenario, habitat quality improves significantly in northeastern Hunan, southeastern Hubei, northern Jiangxi, southern Anhui, Shanghai, eastern Guangxi, southern Guangdong, and southwestern Heilongjiang from 2025 to 2030. Under the ecological conservation scenario, habitat quality gradually improves in southern Guangdong, Chongqing, southwestern and northeastern Shandong, northeastern Anhui, northeastern and southwestern Heilongjiang, and northwestern Jilin. Under the integration of culture and tourism scenario, habitat quality gradually decreases in eastern Nei Mongol, Shanghai, and central-western Xinjiang.

3. **Analysis of the Relationship Between Ecosystem Services and Human Activity Intensity in China's Waterfront Areas**

From 2025 to 2030, the relationship between ecosystem services and human activity intensity in China's waterfront areas under three scenarios shows insignificant changes, with variations mainly concentrated in southern Guangdong. The high-quality ecosystem service areas under all three scenarios are concentrated in western Yunnan, southern Sichuan,

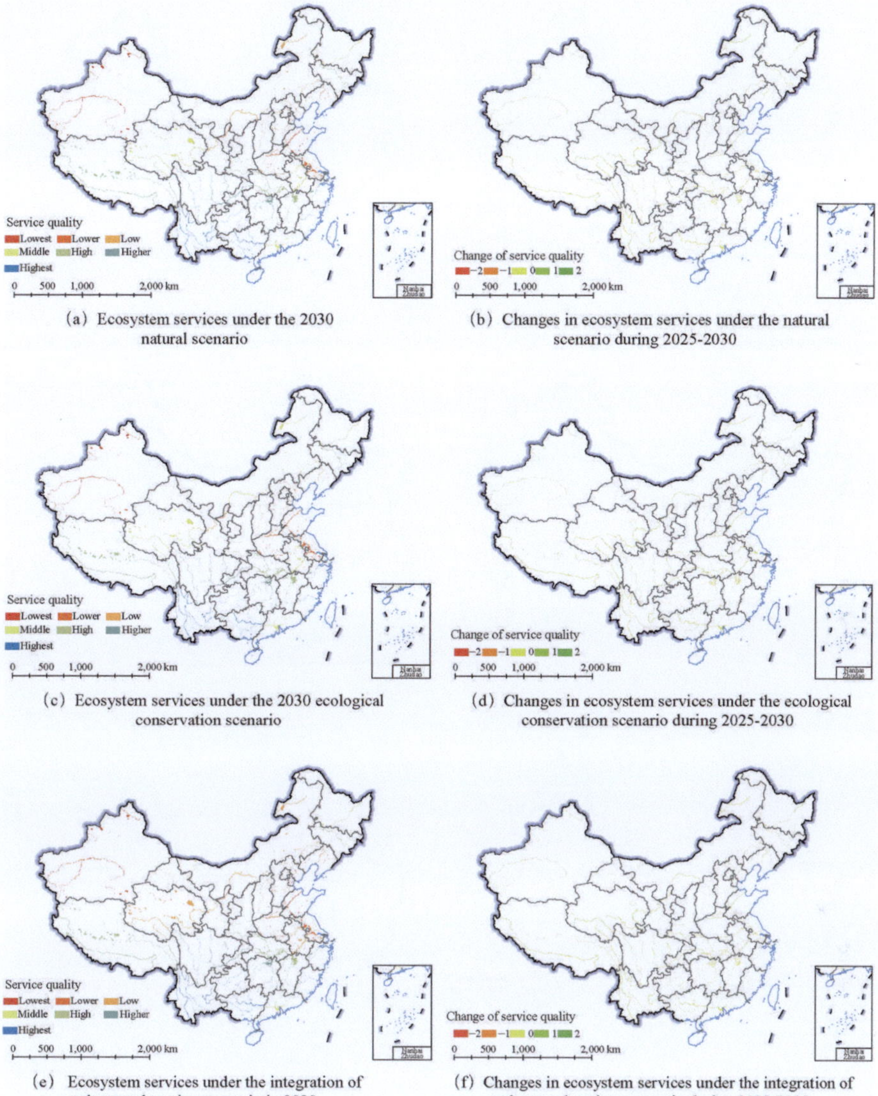

(a) Ecosystem services under the 2030 natural scenario

(b) Changes in ecosystem services under the natural scenario during 2025-2030

(c) Ecosystem services under the 2030 ecological conservation scenario

(d) Changes in ecosystem services under the ecological conservation scenario during 2025-2030

(e) Ecosystem services under the integration of culture and tourism scenario in 2030

(f) Changes in ecosystem services under the integration of culture andtourism scenario during 2025-2030

Fig. 9.5 Changes in ecosystem services under different land use scenarios in China's waterfront areas from 2025 to 2030

northern Guangxi, and eastern Fujian, where ecosystem services are good and the ecological environment is excellent. Meanwhile, the results of the relationship between ecosystem services and human activity intensity in these areas are all extremely high, indicating that the supply of ecosystem services far exceeds human activity intensity. Therefore, it is advisable to appropriately increase human activity intensity, enhance waterfront area tourism development and construction efforts, and promote the development of waterfront area recreation. However, in areas such as southern Beijing, Tianjin, northeastern Hebei, northwestern Shandong, northeastern Henan, western Jiangsu, northern Ningxia, and western Xinjiang, the results of the relationship between ecosystem services and human activity intensity are all extremely low, suggesting that human activity intensity far surpasses the supply of ecosystem services. Therefore, it is necessary to strengthen ecological conservation and restoration, reduce

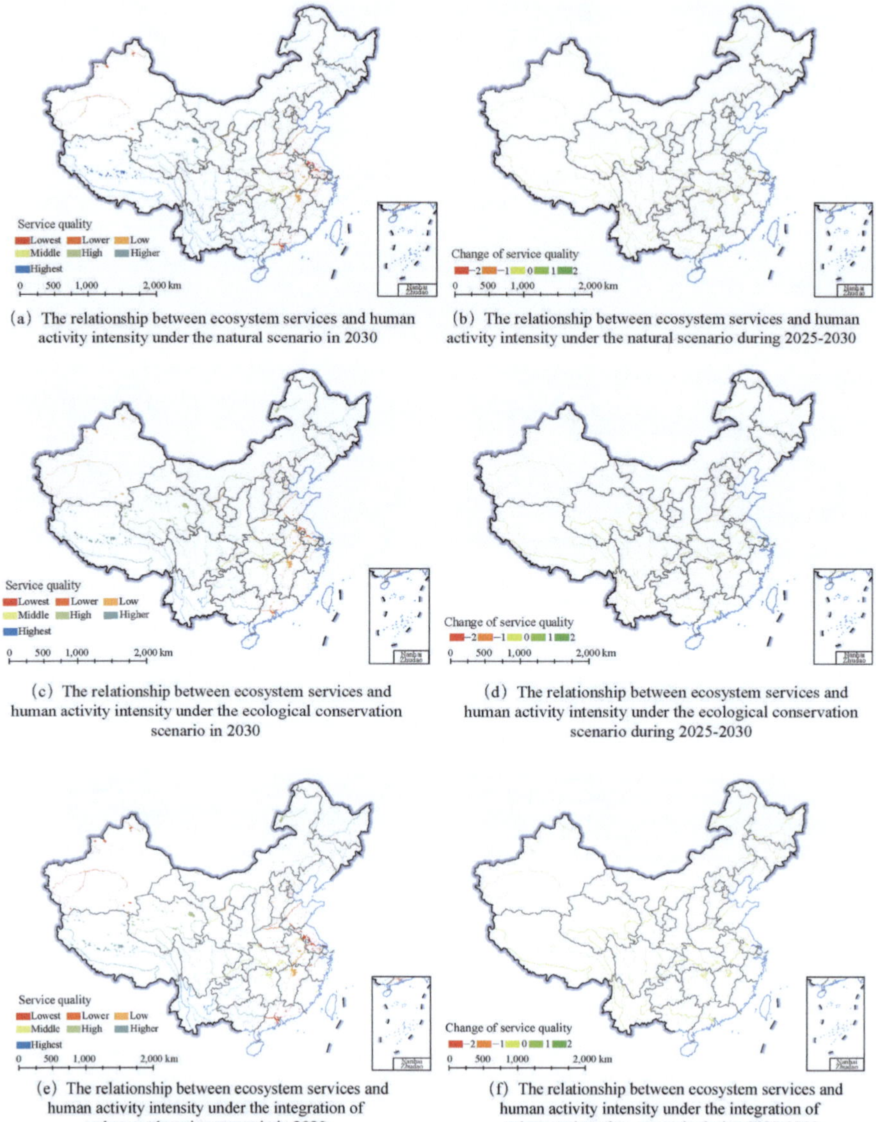

(a) The relationship between ecosystem services and human activity intensity under the natural scenario in 2030

(b) The relationship between ecosystem services and human activity intensity under the natural scenario during 2025-2030

(c) The relationship between ecosystem services and human activity intensity under the ecological conservation scenario in 2030

(d) The relationship between ecosystem services and human activity intensity under the ecological conservation scenario during 2025-2030

(e) The relationship between ecosystem services and human activity intensity under the integration of culture and tourism scenario in 2030

(f) The relationship between ecosystem services and human activity intensity under the integration of culture and tourism scenario during 2025-2030

Fig. 9.6 Relationship between ecosystem services and human activity intensity in China's waterfront areas under different scenarios from 2025 to 2030

recreational activities, appropriately decrease human activity intensity, and promote sustainable development in waterfront areas (Fig. 9.6).

Highlights

- The InVEST model for ecosystem services, RStudio model, PLUS model for land use evolution, and GIS spatial analysis techniques were comprehensively used to explore the trade-offs and synergies between ecosystem service functions and human activities in global waterfront areas from 2010 to 2020, and analyze spatial differences in China's waterfront areas under different development scenarios in the future (2025 and 2030), providing auxiliary support for decision-making to better achieve the SDGs in global waterfront areas.

- **Discussion and Outlook**

Based on four indicators of SDG 6.6, 8.9, 11.3, and 15.1, this case study utilized satellite remote sensing products, OSM data, environmental data, and related human activity data to complete the dataset on the trade-offs and synergies between ecosystem services and human activities in global waterfront areas (2010, 2015, 2020) from a waterfront dimension and a multi-indicator intersection perspective, and simulated and predicted China's waterfront areas. Based on the analysis results, recommendations are made to strengthen ecological restoration and reduce human activity intensity in waterfront areas in southern Beijing, Tianjin, northeastern Hebei, northwestern Shandong, northeastern Henan, western Jiangsu, northern Ningxia, and western Xinjiang, to promote sustainable development and the construction of waterfront tourism. This case study demonstrates the powerful advantages and application potential of spatialization methods for environmental remote sensing big data, with the potential to integrate quantitative results on the trade-offs and synergies between ecosystem services and human activities into online platform systems for digitization, specialization, and operationalization. In accordance with China's BRI and the concept of building a community with a shared future for humanity, this report is expected to promote global attention to the trade-offs and synergies between ecosystem services and human activities in waterfront areas, drive international cooperation, enhance people's well-being, and provide information support for the sustainable development of global waterfront areas and the high-quality construction of China's waterfront areas.

9.3.3 Spatiotemporal Analysis of Economic Development Equilibrium based on Big Earth Data

Target: SDG 10—Reduce inequality within and among countries.

- **Background**

Since the 1980s, the negative impacts of economic globalization and liberalization, coupled with the uneven distribution of income arising from the transition of socialist countries, have ignited extensive debates (Man and Cao 2021; Smithers et al. 2016). Disparities in regional development pose a significant threat to national security, social stability, public well-being, and economic sustainable development. Promoting balanced regional development and narrowing regional gaps have become a global consensus (Karim 2023). The SDGs regard the reduction of inequality within and among countries as a core objective. Since the inception of the BRI, international society has shown widespread interest and active support for strengthening regional cooperation and coordination. This initiative not only inaugurates and shapes a new phase of globalization, but also serves as an accelerator for achieving the SDGs (Liu et al. 2018). This case study aims to monitor evolving patterns of regional equilibrium under the framework of the BRI, and to summarize the role of this initiative in advancing balanced regional development.

Most economic and social statistics published by governments and international organizations are reported at the country level, posing limitations for multi-scale analyses of global economic disparities. Traditional economic datasets often exhibit inconsistencies in statistical definitions or missing data across different time periods, making them difficult to compile. Consequently, previous global studies on economic equilibrium frequently suffer from geographical omissions, focusing on a particular region or single country, or merely relying on the country level as the unit of analysis. However, studies based solely on national-level data may underestimate the actual degree of economic inequality. Moreover, global or cross-regional examinations of economic development equilibrium remain relatively scarce. By contrast, NTL remote sensing imagery effectively bridges these gaps by providing consistent large-scale economic data (Burke et al. 2021).

In addition, the BRI underscores connectivity as a strategic pivot to facilitate the free and efficient flow of production factors. Given its high degree of internationalization, rapid speed, and suitability for long-distance travel, air transportation has played a pivotal role in driving the region involved in the BRI. The accessibility and reliability of aviation networks are crucial determinants of air transportation efficiency. Therefore, the second objective of this case study is to investigate air transport patterns within the region involved in the BRI and, from the perspective of air transport, to propose recommendations for promoting balanced economic development.

- **Data**
  - NTL remote sensing imagery. This case study emploled a globally calibrated NTL dataset (2001–2020) synthesized by other researchers. By integrating sub-national boundaries and international borders in ArcGIS, NTL brightness values were extracted for each country and each province.
  - Population density data. The population dataset utilized in this case study was the global population distribution dataset provided by Oak Ridge National Laboratory in the United States (https://landscan.ornl.gov/).
  - Direct flight data. The number of direct flights during the final week of August 2018 and November 2022 was obtained through Google Flight and cross-referenced using Qunar and Skyscanner.

- **Method**

This case study constructed a three-tier structure encompassing regional, national, and sub-national levels, thereby enabling the derivation of economic disparity patterns at both regional and national scales. Specifically, regional-level economic disparities were calculated based on national-level data, allowing for a distinction between intra-regional and inter-regional differences. Meanwhile, national-level economic disparities were derived from provincial-level data, categorized inter-country and intra-country differences. The Theil decomposition method is a commonly employed approach in analyzing economic inequality. In addition to examining economic inequality across different hierarchical levels, it can also dissect the influence of various social factors (such as population) on economic inequality. Compared to the Gini coefficient and the coefficient of variation, the Theil decomposition index is widely used to quantify economic disparities across multiple levels. On these grounds, this case study applied the Theil decomposition method to investigate economic disparities. Furthermore, this case study selected the 150 countries that have signed cooperation documents under the BRI with China, thereby defining the region involved in the BRI for the purpose of this analysis.

- **Results and Analysis**

1. **Spatiotemporal Evolution of Economic Equilibrium at the Global Scale**

   (1) Components of global economic inequality. Between 2000 and 2020, global economic inequality consistently showed a downward trend. In terms of its composition, global economic inequality is marked by pronounced regional characteristics (Fig. 9.7).

   (2) Intra-regional economic inequality. Global economic inequality displayed a distinct regional pattern, and this configuration underwent minimal change during the period from 2000 to 2020 (Fig. 9.8).

   (3) Economic inequality within countries. Pronounced disparities existed within country boundaries, and between 2001 and 2020, these disparities underwent significant shifts. In particular, the range of imbalances within the region involved in the BRI became more concentrated, warranting a focused examination of economic disparities in this area (Fig. 9.9).

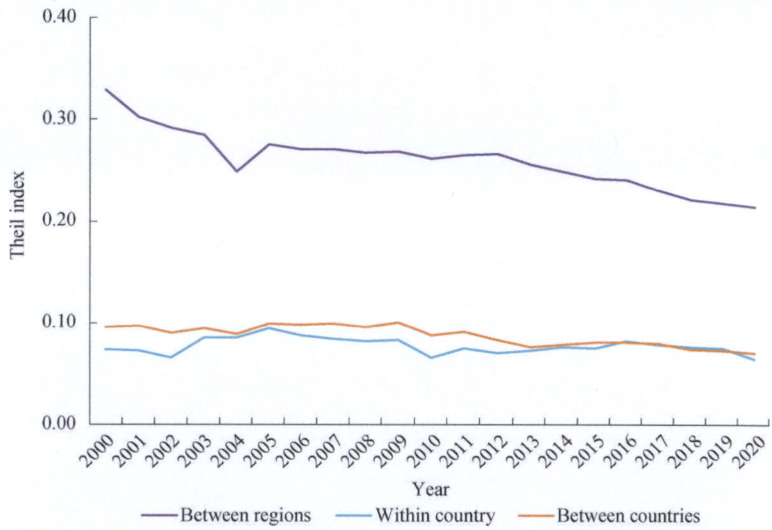

Fig. 9.7 Composition of the global Theil index (2000–2020)

2. **Spatiotemporal Evolution of Economic Equilibrium in the Region Involved in the BRI**

The findings indicate that: (1) Before 2013, the imbalance index among the region involved in the BRI countries fluctuated. After 2013, however, the index showed a stable downward trend (the green line in Fig. 9.10). (2) Economic inequality in the region involved in the BRI is mainly driven by inter-regional disparities. Although the index has been declining over time, its share within the total Theil index has continuously risen. Inter-country inequalities rank second, and intra-country inequalities are minimal.

A further comparison between economic inequality decline at the global scale and that in the region involved in the BRI reveals that the global rate of decrease was consistently higher prior to 2013. After 2013, however, the reduction in economic inequality within the region involved in the BRI surpassed the global average, shifting the advantage to the region involved in the BRI. For instance, from 2013 to 2016, global economic inequality declined by 1.01%, whereas the overall inequality among the region involved in the BRI dropped by 1.67%. This outcome highlights the positive contribution of the BRI to promoting deeper regional and international integration. Its emphasis on "common development", "cooperative development", and "equal development" provides a practical pathway for realizing the UN SDGs, thereby contributing to a valuable Chinese perspective (Fig. 9.11).

3. **Evolution of the Aviation Network Pattern in the Region Involved in the BRI**

The spatial distribution of the aviation network was analyzed, and the results indicate the following: (1) The aviation network within the region involved in the BRI exhibited significant inequality across three aspects: airport layout, route distribution, and weekly flight frequencies. (2) Aviation connections in the region involved in the BRI were predominantly intra-regional, with flights within-regions accounting for 81.67% and 83.36% of the total in 2018 and 2022, respectively. (3) The East Asia region had the highest share of internal connections, with frequent flight interactions concentrated among cities in China's eastern coastal areas (Table 9.2).

Overall, there are several shared characteristics between air transport and the distribution of economic equilibrium. First, both display spatial similarities in the geographical distribution of regional economic inequality. Second, both exhibit relatively high levels of regional disparity.

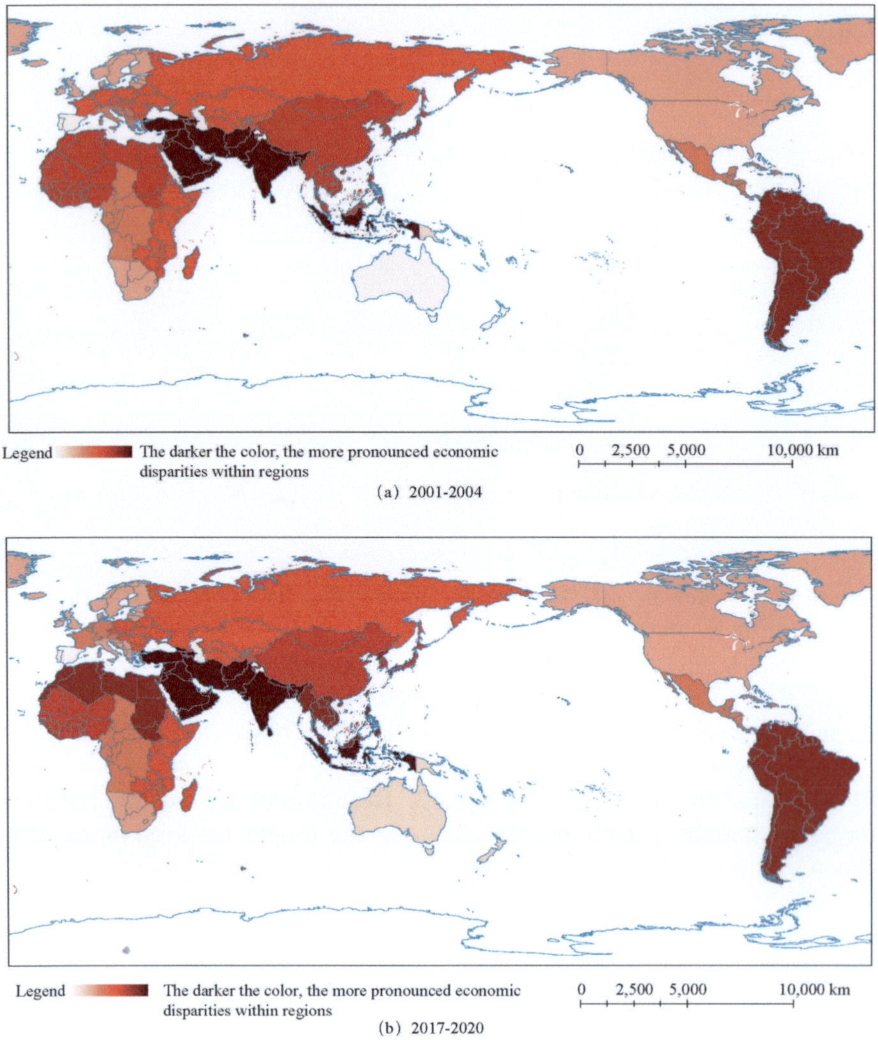

Fig. 9.8 The economic inequalities within-regions worldwide (2001–2004, 2017–2020)

Numerous studies have demonstrated an interdependent relationship between economic development and air transport. On the one hand, air transport serves as a key driver of economic growth, as imbalances in transportation infrastructure are a primary source of regional economic disparities. On the other hand, because air transport is a high-cost mode of conveyance, its development is substantially propelled by economic growth itself. Accordingly, the construction of aviation infrastructure can mitigate economic disparities in two principal ways. The first involves locational factors: building aviation infrastructure enhances a region's accessibility and can even strengthen its locational advantages. In areas disadvantaged by poor locational conditions, investing extensively in aviation infrastructure can effectively upgrade the region's locational profile. The second involves increasing infrastructure investment, which in turn boosts a region's overall influence. Through improvements to locational attributes and the infusion of capital, aviation infrastructure construction contributes to narrowing regional economic gaps.

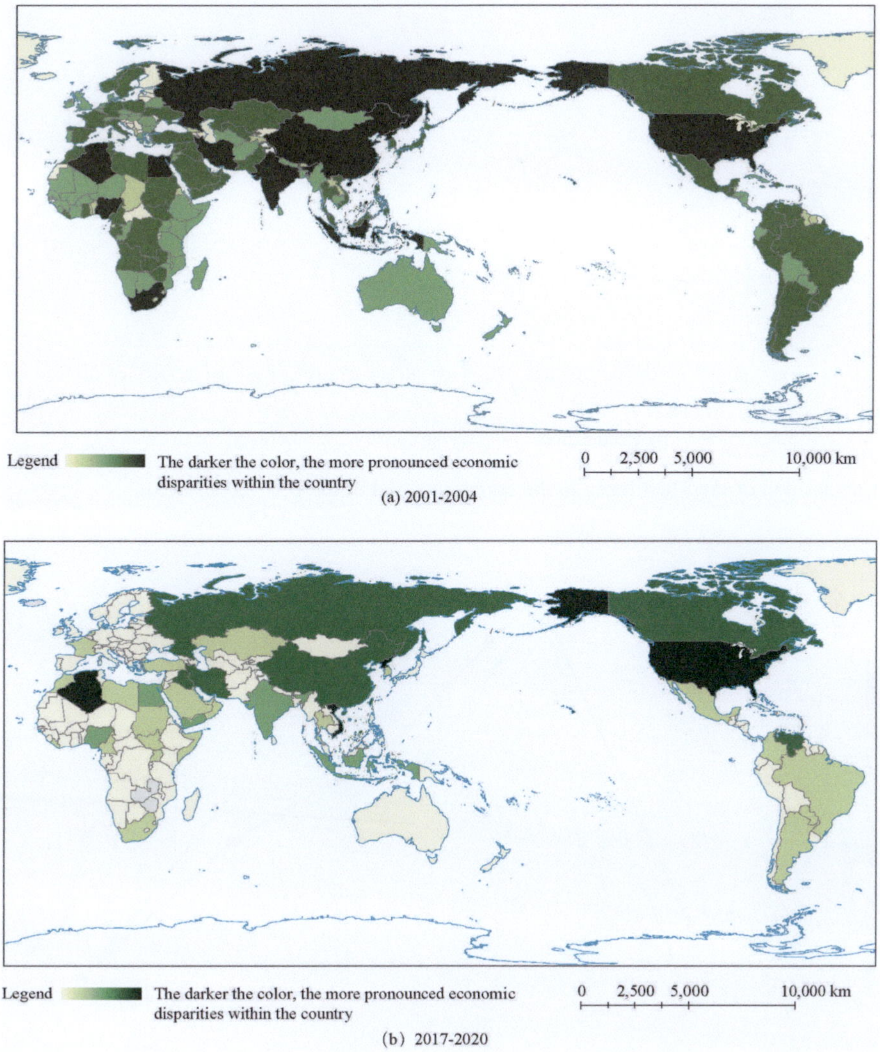

Fig. 9.9 The economic inequalities within-countries worldwide (2001–2004, 2017–2020)

Highlights

- By employing NTL remote sensing imagery, this case study produced per capita economic development data at various spatial scales worldwide, thereby capturing changes in global economic inequality from 2000 to 2020.
- Global economic inequality showed a steady decline between 2000 and 2020, yet exhibited pronounced regional characteristics. These entrenched regional disparities proved difficult to overcome.
- For the first time, data were used to validate the role of the BRI in addressing SDG 10—namely, reduce inequality within and among countries. The findings indicate that this initiative offers a Chinese solution to the challenge of mitigating inter-regional inequalities.

- **Discussion and Outlook**

Using NTL remote sensing imagery, this case study generated per capita economic development data for 2000–2020 on a global

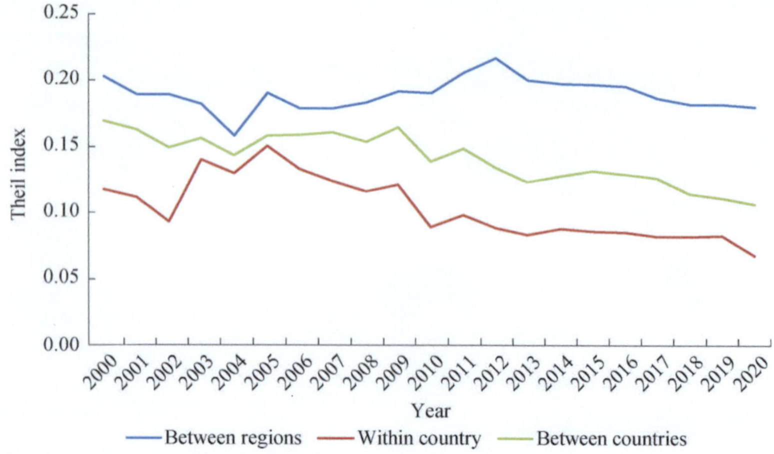

Fig. 9.10 Composition of the Theil index in the region involved in the BRI (2000–2020)

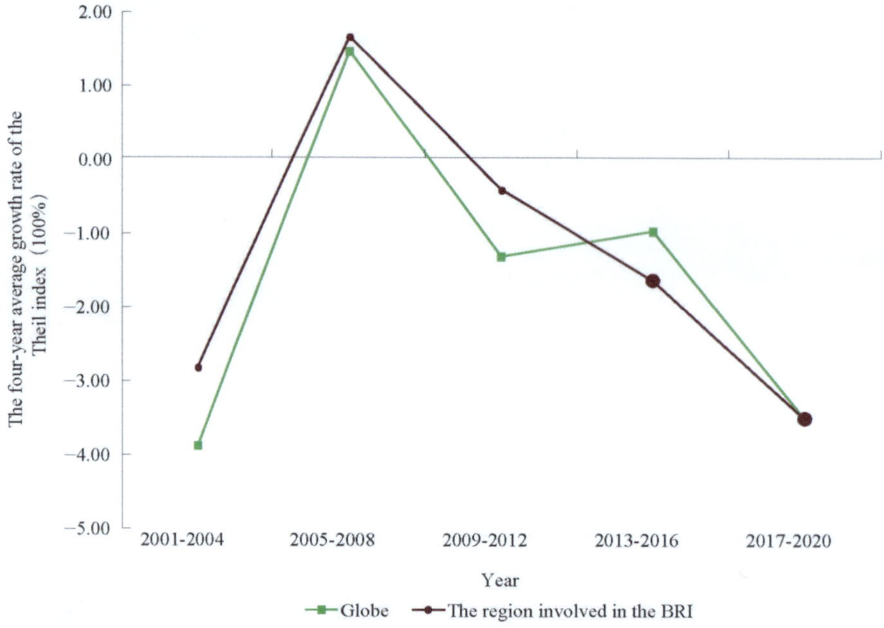

Fig. 9.11 The decline in economic inequality globally and the region involved in the BRI

scale. Adopting a three-tier analytical framework—encompassing the regional, national, and provincial levels—and applying the Theil index alongside its multi-stage decomposition, we separately investigated the evolution of economic disparities at the intra-regional, inter-country, and intra-country scales worldwide. We also compared economic inequality before and after the introduction of the BRI. The results are as follows: (1) Global economic inequality showed a sustained downward trend from 2000 to 2020. (2) Economic inequality worldwide manifested strong regional characteristics, and these entrenched inter-regional disparities are difficult to break down. The BRI provides a Chinese strategy to tackle such inter-regional ineq. (3) Before the BRI, global economic inequality declined at an average